BECOMING SALMON

CALIFORNIA STUDIES IN FOOD AND CULTURE
Darra Goldstein, Editor

BECOMING SALMON

Aquaculture and the
Domestication of a Fish

Marianne Elisabeth Lien

UNIVERSITY OF CALIFORNIA PRESS

University of California Press, one of the most distinguished university presses in the United States, enriches lives around the world by advancing scholarship in the humanities, social sciences, and natural sciences. Its activities are supported by the UC Press Foundation and by philanthropic contributions from individuals and institutions. For more information, visit www.ucpress.edu.

University of California Press
Oakland, California

© 2015 by The Regents of the University of California

Library of Congress Cataloging-in-Publication Data

Lien, Marianne E., author.
　Becoming salmon : aquaculture and the domestication of a fish / Marianne Elisabeth Lien.
　　　pages　cm.— (California studies in food and culture ; 55)
　Includes bibliographical references and index.
　ISBN 978-0-520-28056-4 (cloth) — ISBN 978-0-520-28057-1 (pbk. : alk. paper) — ISBN 978-0-520-96183-8 (ebook)
　1. Salmon farming.　2. Salmon farming—Social aspects.　I. Title. II. Series: California studies in food and culture ; 55.
　SH167.S17L54　2015
　639.3′755—dc23　　　　　　　　　　　　　　　　　　　　2014048458

Manufactured in the United States of America

24　23　22　21　20　19　18　17　16　15
10　9　8　7　6　5　4　3　2　1

In keeping with a commitment to support environmentally responsible and sustainable printing practices, UC Press has printed this book on Natures Natural, a fiber that contains 30% post-consumer waste and meets the minimum requirements of ANSI/NISO Z39.48–1992 (R 1997) (*Permanence of Paper*).

In memory of Asbjørn Lien
(1926–2012)

Once upon a time a salmon made her way upriver. She was exhausted after traveling for months, and hungry too. And just as she buried her eggs in the gravel and felt that familiar urge to let herself go with the flow, back towards the Atlantic Ocean, in the far distance where food was abundant, she had an idea. She thought: "What if we didn't have to do this? What if food was right here, near the shore? What if we could train some other species to get it for us so that we could stay put, right here, and just eat? Wouldn't that be wonderful!"

"Careful what you wish for," said her salmon friend. "You never know—sometimes your wishes come true."

CONTENTS

Illustrations — ix
Acknowledgments — xi

1. Introduction: Salmon in the Making — *1*
2. Tracking Salmon — *27*
3. Becoming Hungry: Introducing the Salmon Domus — *48*
4. Becoming Biomass: Appetite, Numbers, and Managerial Control — *76*
5. Becoming Scalable: Speed, Feed, and Temporal Alignments — *104*
6. Becoming Sentient: Choreographies of Caring and Killing — *126*
7. Becoming Alien: Back to the River — *148*
8. Tails — *164*

Notes — *173*
References — *197*
Index — *209*

ILLUSTRATIONS

FIGURES

1. Vidarøy from the boat *50*
2. Wrasse cover *55*
3. Cages with salmon jumping *58*
4. View from gantry *68*
5. Salmon samples for color analysis *95*
6. Farmed salmon production in Norway, total volume, 1976–2011 *106*
7. Will they eat? Alevins ready for their first feeding *113*
8. The black roof at the smolt production site *117*
9. Distinctions multiply: An overview of practices to differentiate wild, cultivated, and escaped farmed salmon *156*
10. Smolt screw *157*

TABLES

1. Excerpt from monthly report, Frøystad smolt production site *89*
2. Origins of fishmeal in feed pellets *121*

ACKNOWLEDGMENTS

When I first visited a Tasmanian salmon hatchery in November 2001, I had no idea that this was the beginning of an ethnographic journey with salmon that would last for more than ten years. But ethnographic encounters and academic exchanges triggered new questions and took me along unexpected paths. I am grateful for all the encounters—human and nonhuman—that have made this journey so exciting and for the insight, trust, and knowledge that have been granted me along the way.

I am indebted to Richard and Annalisa Doedens in Tasmania for their friendship and for letting me into their business at a time when I could offer very little in return. Trine Karlsrud, Vicky Wadley, and Guy Westbrook were important guides to salmon farming in Tasmania, and I am grateful to them for taking the time.

I wish to acknowledge financial support from the Norwegian Research Council for funding my extended fieldwork on salmon, first in Tasmania through the project "Transnational Flows of Concepts and Substances" and later in Norway through the project "Newcomers to the Farm: Atlantic Salmon between the Wild and the Industrial." The Department of Social Anthropology at the University of Oslo has provided periodic funding for seminars and travels.

This book would not have been possible without access to the salmon industry. I am deeply grateful to the Norwegian company Sjølaks (a pseudonym) for kindly letting John Law and me locate our study on its sites and for its additional generous practical support, including free accommodation at its premises. I would like to thank all those who work for Sjølaks for their warm welcome, their openness, and their help, including their willingness to teach us simple tasks and to let us watch them at work. I hope that my account does justice to their knowledge and their efforts. I am also grateful for additional interviews and observations with many other firms, agencies, and individuals affiliated with the salmon industry, both in Norway and in Tasmania. Most interviews were conducted in confidentiality, and the names of interviewees are withheld by mutual agreement

John Law, Gro B. Ween, and Kristin Asdal were project collaborators through this journey. John Law was an important source of inspiration through his own writings in Science and Technology Studies (STS). Later he became my collaborator in the field, project partner, coauthor, and friend. I deeply value his sense of humor, generosity, intellectual encouragement, and critiques. Much of what I know about STS, I know through him. Most of what I know about salmon in Norway we learned together. This book is indeed an outcome of our ongoing collaboration. While John and I studied salmon aquaculture, Gro B. Ween studied salmon "in the wild." Her fieldwork along the Tana River provides a counterpoint to this book and has continually reminded me that there is much more to salmon than aquaculture. I am grateful for her enthusiasm, intellectual curiosity, and cheerful spirits. Kristin Asdal studied the domestication of cod, and I am grateful for her patient insistence that farmed fish are enacted at many different sites, including in archives and in written reports.

"Newcomers to the Farm: Atlantic Salmon between the Wild and the Industrial" offered a stimulating arena for salmon scholars of many kinds. I am grateful to Bjørn Barlaup, Ben Colombi, Børge Damsgård, Sunil Kadri, and Cecilie Mejdell for sharing their insights, and to master students Line Dalheim, Anita Nordeide, and Merete Ødegård for important contributions to our project. I have also benefitted from the work of other "salmon scholars." Heather Swanson has been an invaluable source of insight and encouragement and has shaped my ideas in significant ways. I am also indebted to Karen Hébert for reading drafts and offering feedback, and to Anne Magnusson, whose fieldwork on aquaculture in Norway preceded mine.

My approach to salmon took a new direction after a Wenner-Gren symposium in 2004, "Where the Wild Things Are Now: Domestication Reconsidered." I am grateful to Rebecca Cassidy and Molly Mullin for inviting me to consider salmon aquaculture in relation to domestication and for convincing me that I would have something to contribute. Later, discussions with David Anderson, who works on Arctic domestication, have inspired me to challenge the conventional approach to domestication and to think differently about the Domus in anthropology and elsewhere.

I also wish to acknowledge all of those who contributed to "Performing Nature at World's Ends," a special issue of *Ethnos* in 2011, and who inspired and provided ethnographic material for comparative analysis between Norway and Australia. Special thanks to Simone Abram, who helped realize a coedited publication and thus a theoretical platform for further analysis, and to Aidan Davison, Adrian Franklin, Stephanie Lavau, David Trigger, Helen Verran, and Gro B. Ween for intellectual support throughout. My work has been shaped by numerous other workshops and conferences, and I am grateful to audiences as well as opportunities to talk about salmon in many parts of the world, including departments or schools in Hobart, Melbourne, Sydney, Aberdeen, Santa Fe, UC Santa Cruz, UC Davis, Lancaster, Stockholm, Linköping, Copenhagen, Aarhus, Tromsø, Trondheim, Bergen, and Oslo. I have also received valuable comments at anthropology and social science conferences—such as the EASA (Bristol 2006), ASA (Auckland 2008), NAT (Stockholm 2012), AAA (San Francisco 2012), CRESC (London 2013), and IUAES (Manchester 2013)—as well as aquaculture conferences in Faro (BENEFISH 2009), Tromsø (Havbrukskonferansen 2008), and Trondheim (Aquanor 2013) and at Stord (Aqkva 2012).

After many years of fieldwork and writing shorter pieces, my book began to take shape when in 2012–13 I had the opportunity to spend a year at University of California Santa Cruz, Department of Anthropology. I can hardly imagine a more stimulating environment. My colleagues in Santa Cruz are role models for cultivating curiosity and intellectual courage, and I am forever grateful to each and every one for inviting me to take part in their academic life. Special thanks go to Andrew Mathews, Heather Swanson, and Anna Tsing for continuous sharing of ideas, "forest conversations," and for constructive comments on early drafts. I am also grateful to Nancy Chen, Shelly Errington, Donna Haraway, Diane Gifford-Gonzalez, and Danilyn Rutherford for their interest and valuable comments, and to graduate students Rachel Cypher, Pierre du Plessis, and Katy Overstreet for excellent questions and discussions.

I have benefitted from valuable support, encouragement, and comments from many other people, including Filippo Bertoni, Lotta Björklund Larsen, Mario Blaser, Marisol de la Cadena, Arturo Escobar, Frida Hastrup, Britt Kramvig, Tess Lea, Francis Lee, Anita Maurstad, Maria Nakhshina, Ben Orlove, Gísli Pálsson, Elspeth Probyn, Vicky Singleton, Paige West, Richard Wilk, and Brit Ross Winthereik.

At the Department of Social Anthropology, University of Oslo, I am surrounded by supportive colleagues and inspiring students. I wish to acknowledge my debt to all colleagues in Oslo and give a special thanks to those who have engaged directly with my work on salmon: Harald Beyer Broch, Rune Flikke, Nina Haslie, Signe Howell, María Guzmán-Gallegos, Marit Melhuus, Knut C. Myhre, Anne-Katrine Norbye, Knut Nustad, Jon Rasmus Nyquist, Jon Henrik Ziegler Remme, Stine Rybråten, Cecilia Salinas, Astrid Stensrud, and Christian Sørhaug.

Jacob Hilton gathered material on Alaskan salmon; Kristian Sandbekk Norsted compiled my endnote library; Jennifer Shontz prepared graphic illustrations; John Law took the many photos for our project and kindly let me use his for many illustrations; and Susanna J. Sturgis and Barbara Armentrout edited the manuscript. At the University of California Press, Kate Marshall, Stacy Eisenstark, Jessica Ling, and Dore Brown have enthusiastically guided me through various stages of publishing, and I am grateful for their support.

My family is a continuous source of gratitude and joy. I am thankful to Vidar, Thomas, Torbjørn, and Eira, and to my mother, Helle, for taking part in this journey and to my husband, Eivind, for his presence in my life. Special thanks go to all our children for embracing Tasmania at an early age, to Eira for her courage to attend schools in Hordaland and join me temporarily during fieldwork, to Eira and Eivind for traveling with me to Santa Cruz for a whole year, and to Malin and Andrés for being part of our family too. Eivind has generously engaged with this book in all sorts of ways and is a continuous source of inspiration and support. My father, Asbjørn, passed away before this book was finished. I am forever grateful for his love and intellectual encouragement and for sharing his passion for writing. I dedicate this book to him.

1

INTRODUCTION

Salmon in the Making

Everything there is, launched in the current of time, has a trajectory of becoming.
—TIM INGOLD, *BEING ALIVE: ESSAYS ON KNOWLEDGE AND DESCRIPTION*

Salmon have come a long way. They were here before us. They were the backbone of coastal human livelihoods, a reliable and seasonal resource, a conduit of nutrients from the depths of distant oceans, conveniently delivered where the land meets the sea. Our history together goes way back. But recently it took off in a new direction.

This book is an ethnography of a "city of fish." It tells the story of how salmon became a husbandry animal and how sentience was extended to salmon. It is about an emerging industry that expanded beyond all predictions and caused a massive rechanneling of marine resources worldwide. It is also the story of how new forms of biocapital transformed the relation of human to nonhuman, of culture to nature, as salmon went global. But most of all, it is a study of novel and unfamiliar encounters between salmon and their people, on and off salmon farms in western Norway, upriver and beyond.

We are in the middle of the fjord. The glacier above the eastern shore reflects the morning sun in shades of pink. The water surface is occasionally broken by a silvery shadow that jumps through the air, a silent reminder that we are not alone. Then a shattering sound breaks the silence, like a hailstorm above our heads. It is the steady hammering of feed pellets through plastic pipes. Half a million salmon will soon be fed. Our working day has just begun, and so has theirs. Farmed Atlantic salmon are bred to be hungry, and their job description is simple: putting on weight.

Cospecies histories are difficult to tell. This is not only because some of us have been trained to think that having a history is a privilege of humans, and thus we have ignored the ways in which nonhuman species have histories too. It is also because we are trained to tell the stories separately. We tend to think of histories as either *their* stories, animal stories, as they unfold without human interference, or *our* stories, with humans cast as the key actors and animals figuring as prey, property, or symbols. *Their* stories find an audience among biologists, *our* stories among anthropologists.

But storytelling practices are changing. Biologists and environmental ecologists are beginning to write humans into their stories of nature. Geologists have launched "Anthropocene" as the label for our time, to bring attention to lasting human impact on the atmosphere and the earth itself. Social and cultural anthropologists are warming up to the idea that the term *social* might include nonhumans too, and they are beginning to explore the ethnographic implications of this possibility.

Archeologists, who never forgot how closely human history is intertwined with the histories of animals and plants, have begun to pay more attention to this interface, with a renewed focus on mutualism and coevolution. But even archeologists struggle when it comes to fish, whose traces are rarely visible. While domesticated goats and sheep reveal glimpses of our shared evolution as pieces of bone in the ground, fish leave few such remains. Ancient middens can tell us about human marine diets but very little about the shape of the fish and how we might have evolved together.

Salmon farming can be seen as the most recent turn in the human history of animal domestication. For the first time, fish are enrolled in regimes of intensive industrial-scale aquaculture. Spearheading the "blue revolution," the unprecedented expansion of salmon aquaculture adds pressure on marine fisheries through an increased demand for fish resources in the North Atlantic as well as the South Pacific. But the emergence of salmon farming is more than a shifting pattern of resource exploitation: as fish are enrolled as husbandry animals, conventional distinctions between fish and animals are also reconfigured. Appropriated by industrial regimes of global food production, farmed salmon have become not only biomass and objects of capital investment but also sentient beings, capable of feeling pain and subject to animal welfare legislation.

Freshwater experiments with trout and salmon propagation have been done for generations, on both sides of the North Atlantic. However, it was not until the early 1970s that experiments with saltwater salmon aquaculture succeeded, and commercial salmon farming as we know it was invented: soon, salmon were enrolled in regimes of intensive production, and the sea was no longer a place only for fishermen, but for farmers too. Agricultural principles were extended to the marine; aquaculture was expanded to the estuaries and the sea. During the 1980s and 1990s salmon aquaculture intensified and expanded to the Southern Hemisphere, while production increased manifold.

. . .

This book explores the recent journeys of these "newcomers to the farm" as well as the novel regimes of animal husbandry in which they take part. I have traced salmon and their people on and off salmon farms, from the "cities of fish" in Norwegian fjords to the dark and humid production rooms where fragile orange eggs are about to hatch. I have followed their trajectory from the North Atlantic to the Southern Hemisphere and explored farmed salmon through paper trails, in research journals, at websites and international business exhibits. I have followed them up along rivers in western Norway, to where their distant cousins still breed, guided by salmon anglers and biologists whose concern about the wild salmon stock makes escapees from the salmon farms unwanted intruders. But most of the time, I have worked with people for whom feeding, processing, and caring for farmed salmon is a livelihood that they value and are deeply engaged in.

Unfamiliar encounters? Yes indeed. Fish are cold. They live in water. They are mostly out of sight. They are silent. Their staring eyes show no visible emotion. Their body language is difficult to interpret. All of this limits the cues that humans can respond to. And yet our response is crucial. Domestication is a two-way process. Ever since our ancestors began to share their households with animals, husbandry relations have been a mutual affair. This book is about how husbandry practices are now extended to fish. It is about a recent journey on a steep learning curve. But it is also about remote communities of Norwegian farmer-fishermen who have found themselves at the forefront of global aquaculture, shipping truckloads of fresh salmon to Paris, Tokyo, Moscow, and Dubai.

Domestication is often told as a fundamental shift in the way humans engaged with animals and plants, with control and confinement as key modes of ordering human-animal relations. Guided by the water world of salmon and their caretakers, this book challenges that approach and suggests that *mutuality, uncertainty,* and *tinkering* are better terms with which to capture the productive entanglements of humans and their surroundings. Rather than model aquaculture on an outdated narrative of human progress as "control of nature," I invite the reader to take part in exploring the novel opportunities that more-than-human entanglements offer, as well as the risks that they entail. In that sense, this book is a mutual interrogation of what salmon farming and domestication are, and what they might become.

NOVEL ENTANGLEMENTS

Atlantic and Pacific salmon have sustained communities along the North Atlantic and the North Pacific rims for as long as there are historical records and probably much longer. In many instances, humans have also intervened in salmon trajectories through material arrangements that divert or delay their journeys upstream. They have enhanced the salmon's reproductive capacity by extracting roe and milt from broodstock and distributing fertilized eggs across entire riversheds, and they have unintentionally altered the genetic profile of their most sought-after local salmon stocks through overfishing and selective predatory practices.[1] In these and many other ways, human lives and salmon lives have nearly always been intertwined, and much of what we see in salmon farming today could be seen as continuities, or instances of intensified entanglements. And yet, there are also ways in which contemporary salmon farming appears fundamentally new.

One of the most striking features of contemporary salmon farming is its unprecedented scale: salmon aquaculture has quite simply been a huge success, from a business point of view, with a potential for expansion far beyond what anyone could predict. By the late 1990s it was estimated that more than 95 percent of all Atlantic salmon living in the world had been raised at a fish farm (Gross 1998). Since then, the global production has more than doubled and the percentage is likely to have increased. By 2009, aquaculture supplied half of the total fish and shellfish for human consumption. By 2012, aquaculture was among the fastest-growing food-producing sectors in the world (spearheaded by farmed salmon) and expected to eclipse the global production of beef, pork, or poultry by the next decade (FAO 2012; Nærings- og fiskeridepartementet [Ministry of Trade, Industry and Fisheries] 2013). As a result of such massive growth, and because farmed salmon depend on a steady supply of fish meal and fish oil to thrive, the growth of aquaculture adds pressure on global marine fisheries. Wild fish are used to produce feed pellets for terrestrial livestock, and farmed salmon add yet another demand on this scarce resource. Hence, aquaculture's share of the global use of fish oil and fish meal more than tripled from 1992 to 2006 (Naylor et al. 2009, 15103; FAO 2008).[2] As a predator in the marine food chain, farmed salmon have a key role in this global rechanneling of marine resources from human consumption to animal feed and from terrestrial husbandry to marine husbandry.

Domestication lies at the heart of this massive transformation. Yet, while the idea of salmon as "newcomers to the farm" points to a historical shift, domestication of aquatic animals is not entirely new, nor could a model of domestication fully account for the massive global expansion we are currently witnessing. In this book, I mobilize domestication as a comparative tool as well as a conceptual placeholder for more-than-human practices that can be studied ethnographically across various domains. Drawing on recent insight in archeology and anthropology as well as studies of human-animal relations, I approach domestication as sets of relations across species barriers that enable and enact particular biosocial formations, or reproductive practices through which humans and nonhumans mutually inhabit each other's worlds and (intentionally or unintentionally) make space for one another. As they unfold, such relational practices often involve the rearrangement of space, the appropriation of place, or the arrangement of time in cyclical sequences. Consider, for example, seasonal shifts of sowing and harvesting, the appropriation of a field for agricultural purposes, or the movement of and with pastoral herds. Often, they also involve novel infrastructural arrangements, such as a barn, or a chicken coop, or indeed a salmon pen, which—as we shall see—

serve to gather or facilitate particular relational practices. I approach such sites of human-animal encounters as heterogeneous gatherings of human and nonhuman entities that define or enact what a domesticated animal may become. Inspired by the literature on domestication, I shall refer to such sites as domus (see chapter 3), fragile assemblages of beings and things that, as long as they hold together, constitute the conditions of growth and reproduction of humans as well as of nonhuman beings.

Marilyn Strathern has reminded us that it matters what ideas we use to "think other ideas with" (Strathern 1992, 10). Her insight has inspired further twists.[3] To state that "it matters what materials we use to think other materials with" or that "it matters what practices we use to think other practices with" suggests novel ways in which to attend to the lateral dimension of ethnographic practice. In the context of salmon aquaculture, they serve as reminders that there is no obvious context out there waiting to be revealed, no theory providing the obvious analytical anchor for the material at hand, but instead, endless opportunities for association and juxtaposition, each with the potential for taking the analysis in a new direction.

The concept of domestication frames my story, while at the same time offering a comparative edge against which my material can be "thought." The choice is not obvious. There are many other narratives circulating in the realm of aquaculture discourse, ready to eclipse an analysis of salmon farming. "Industrial capitalism," "global food production," "environmental degradation," and "wild salmon extinction" are but a few alternative framings. When I choose to "think salmon through goats and cows," it is not because the alternatives are not also relevant but because domestication offers opportunities to mobilize comparisons that I find particularly intriguing. Most importantly, it helps me decenter, or temporarily suspend, more conventional normative narratives, which tend to come with a ready-made cast of heroes and villains, good and bad. It also allows me to think through aquaculture without being immediately locked into the conceptual dualism of nature and society as oppositional domains,[4] a dichotomy that underpins so many contemporary debates about the environment and the "wild."

Domestication works here as a comparative tool that mobilizes connections and continuities across various temporal and topical domains, while encompassing fundamental historical concerns about how we humans nourish ourselves. In this way, it has guided the analysis in unexpected directions. This has been possible, however, only after a critical reassessment of domestication as a conventional narrative.

DOMESTICATION AND THE NARRATIVES WE LIVE BY

If anthropology is to operate as an ongoing "decolonialization of thought,"[5] then it should involve a constant questioning of narratives we live by. Domestication can be seen as one such narrative, which has sustained, justified, and made legible particular historical trajectories and biosocial relations that are familiar to Euro-Americans and associated with progress and modern civilization. As such, it encompasses ways of life that are now hegemonic on a global scale (sedentary agriculture, private property, coercive husbandry, extraction of natural goods) and that both reproduce and are justified by dualistic oppositions of culture and nature as well as assumptions about human exceptionalism. Control and spatial confinement (of husbandry animals as well as of native peoples) have been invoked as essential elements of such processes. In anthropology, domestication has similarly been associated with human mastery, taming, and control as well as with similar hierarchical and classificatory distinctions (Cassidy and Mullin 2007; Candea 2010).

That salmon and domestication are uncommon companions in the scholarly literature is therefore hardly surprising: whereas the former epitomizes the wild, the latter typically tells the story of how humans conquered nature. The global expansion of salmon farming could have been told as the story of how nature is overtaken as industrial capitalism expands to novel aquatic "terrains,"[6] or a final stage along an evolutionary journey of animal domestication (Zeder 2012).[7] And, indeed, this is how the story has often been told. However, this view of domestication tends to gloss over heterogenic specificity as well as instances of noncoherence that proliferate where particular "forms of life," such as industrial food production, and organic "life-forms," such as its products (feed pellets, farmed salmon fillets), intersect (Law et al. 2014, 173; Helmreich 2009, 6).

To assume that we already know what domestication entails is to run the risk of marginalizing the unpredictable and often unexpected practices and outcomes that unfold when very different worlds are simultaneously performed—worlds that do not necessarily map smoothly onto conventional models of domestication. So, rather than analyzing the recent growth of the salmon farming enterprise as yet another outcome of human mastery over "nature" (or a result of neoliberal capitalism's mindless exploitation of animal beings), I weave the connections between domestication and salmon farming in multiple ways, asking these questions: What are the specific practices and choreographies through which farmed salmon are currently being enacted? What formations of humans and nonhumans emerge as a result, and how do they hold together? As salmon are "becoming with" humans in new ways and through multiple practices, what salmon are enacted in the process? And finally, how can such salmon shed light on processes of domestication?

These questions draw the attention to processes of biosocial becoming as well as to the possibility that salmon may emerge as "more than one and less than many" (Strathern 1991, 36). Hence, the book is organized to draw attention to the partially connected and hardly coherent salmon that emerge through aquaculture practices. *Hungry, biomass, scalable,*[8] *sentient,* and *alien* are a few qualifiers that ground and gather sets of interrelated practices on and off salmon farms and are consequential for what salmon become (see chapters 4–7).

My assumption, then, is that domestication can involve a series of transformative relations through which human and animal are shaping one another. Rather than casting it as a passive figure already locked into a particular trajectory of increasing human control, I approach salmon as an emergent companion species, complicit in, as well as resisting, the various projects in which it is enrolled. Attentive to economics as well as affect, I highlight the fragile and contingent practices that constitute salmon aquaculture and the multiple ways of "becoming salmon" that emerge as a result.

DOMESTICATION: SHIFTING CONFIGURATIONS

For more than a century, the idea of domestication has served as a faithful companion in European practices of creating order and making sense of the world, as well as our place in it. As a powerful image of how humans have transformed their nonhuman surroundings, domestication has justified hierarchies through specific regimes of temporality, which distribute people and places along a single evolutionary trajectory: from the wild to the domesticated. As such, domestication has figured as a particular mode of world-making in which civilization comments on itself.

Domestication can be defined in a number of ways, and its meanings have shifted over time, both inside and outside of academic discourse. What follows is an attempt not to nail it down but rather to open up a few of the conversations in which domestication figures as a key concept or ordering device. I visit biology, archeology, and anthropology, asking not only what domestication is in these epistemic fields, but also what kind of work it does, what relations it enacts and sustains—or, indeed, what relations it glosses over—and what kind of ordering exercises the concept of domestication performs.

"Domestication" is derived from the Latin word *domus,* which in ancient Rome referred to a type of house occupied by the wealthier classes.[9] Dictionary definitions link domestication to hearth and home as well as to the process through which something is either converted to domestic use (tamed) or made to feel at home (naturalized).[10] A process of transformation is central, and the word evokes a distinction between something that is contained within house, household, or home and something that is not (yet) contained within such a setting. A sense of boundary is thus enacted, which may, through the process of domestication, be transcended or overcome.

Popular narratives tend to describe domestication as a unilinear process that originated from a particular point, or cradle, and spread to other parts of the world through diffusion, warfare, or conquest. Domestication is associated with the so-called Neolithic Revolution, which is portrayed as a transformative moment when human beings began to control nature through agricultural and husbandry practices. The term itself denotes a turning point in the history of human civilization (Childe and Clark 1946). In popular narrative, the Neolithic Revolution marked the beginning of an evolutionary process of development, one that is often attributed to human agency and intentionality. The popular TV series *Mankind: The Story of All of Us* (History Channel 2012) is typical example. The first episode, "Inventors," which deals with the shift to agriculture, is introduced as follows: "On a unique planet, a unique species takes its first steps: Mankind begins. But it's a world full of danger. Threatened by extinction, we innovate to survive—discovering fire and farming; building cities and pyramids; inventing trade—and mastering the art of war. From humble beginnings, we become the dominant creature on the planet. Now the future belongs to us."

Archeological excavations have been important in mapping this historical journey. They have taught us that domestication began some 6,000–10,000 years ago in the Middle East and spread from there to other parts of Eurasia and Africa, irreversibly transforming landscapes, societies, and animals and plants on its way (Cassidy and Mullin 2007; Vigne 2011). With domestication came a surplus that allowed, but also depended on, larger human settlements (to herd the animals, till the soil, plant, harvest, and so on).[11] This in turn paved the way for division of labor, social stratification, private property, and state formation: in short, the world as we know it. In some versions, domestication is the mark of evolutionary success, while others emphasize its darker side. But both put domestication to work in order tell *other* stories about humanity, ecology, or the future of our planet.[12] Hence, they locate domestication at the heart of a fundamental rift between the civilized and the savage, the tame and the wild, and culture and nature, with far-reaching consequences for colonial and postcolonial politics, nature management, the institutional structure of scientific research, and technologies of governance.

Control and spatial confinement are often invoked as essential features of domestication. Consider, for example, the much-cited definition of domesticated animals as "bred in captivity for purposes of subsistence or profit, in a human community that maintains complete mastery of its breeding, organization of territory and food supply" (Clutton-Brock 1994, 26). Human control over plants and animals is portrayed as a modern enactment, featuring "Man" as the individualized agent, the human creator, who holds the power to act upon Nature, which is merely acted upon (see, e.g., Descola 2012, 459). With an emphasis on control, mastery, and confinement, domestication provides a model within which specific biosocial relations become legible, normalized, or self-evident. As such, it could be seen as a paradigmatic template, or a factish myth through which a broad range of phenomena is ordered, expressed, or shaped.[13] In this way, domestication has served as a powerful trope for thinking about humanity and for our relations with nonhuman living beings.

At the same time, domestication is indeed a fluid concept, highly mutable as it travels across disciplinary boundaries and from popular to scientific discourse.

Domestication in Biology

Biological definitions draw attention to changes "in the flesh." Domestication is seen as the process whereby a population of animals or plants is changed at the genetic level, accentuating traits that benefit the humans. Similar approaches are applied in relation to biodiversity; the Convention on Biological Diversity (UNEP 1992) defines a domesticated species as a "species in which the evolutionary process has been influenced by humans to meet their needs" (Art. 2). This echoes Charles Darwin, who used the term *domestication* to refer to the transformation of animals from wild species to *recognized breeds* (Leach 2003, 356).[14] Bilateral relations between human and single nonhuman species are emphasized rather than multispecies entanglements, and a notion of human intentionality and agency remains central to the way distinctions are being made.

Traditionally, morphological and, to some extent, behavioral differences were highlighted to distinguish domesticated from nondomesticated species. But with the rapid developments in genetic research, differences are now more often mapped as genetic. In the case of salmon, for example, identifying detectable genetic differences between different strains of salmon in a particular river is required in order to establish whether the salmon in question is in fact "wild" (Lien and Law 2011; Nordeide 2012). Hence, a shift has taken place that locates wildness "in the genes," or in the genotype, rather than in the phenotype or in the landscape as such. In this version, domestication becomes different from taming, which for biologists means particular human-animal encounters during an animal's lifetime, a process that does not necessarily have a genetic impact on the offspring.[15]

With a focus on genetic change, biological definitions of domestication assume some kind of human interference, such as human control over animal breeding, of animal movement, or as selective pressures—intentional or unintentional—in the manmade environment (see Russel 2002).[16] The key question is not so much *why* or *how* it happened but rather the result in terms of genetic differentiation. Hence, the word *domestication* enacts a detectable difference between life-forms that are "pristine" and those that are somehow "invented." With this sorting device, biologists can help to identify the conservation value inherent not only in the messy liveliness *within* salmon rivers but also in entire rivers, or patches of landscape, and they can specify with some precision which ones are "worth looking after" and which ones are not. Because it offers a tool for distinguishing the "wild" from life-forms that are affected by human impact, domestication is at the core of environmental discourse as well and serves as a source of justification for policy decisions in relation to contested alternative land use.

Domestication in Archeology

For archeologists of the Victorian era, domestication was a historical event, a moment of radical transition from one type of society to another (Smith 2001). This image of an abrupt transformation is articulated most clearly by the term *Neolithic Revolution*.[17] The term was coined by archeologist V. Gordon Childe and originally conceived of as a revolutionary change brought about by man. An evolutionary paradigm underpinned this approach, and temporal discontinuity is highlighted. Inspired by visits to Russia as well as by the evolutionary paradigm of L. H. Morgan, Childe adopted the Marxist terms *savagery, barbarism,* and *civilization* to denote stages that were separated first by the Neolithic revolution and then by what he called the Urban revolution (Trigger 1980; see also Childe 1958).[18] Within this approach, domestication was portrayed as the *transition in human food procurement* from hunting and gathering to practices that involve different combinations of agriculture, sedentism, and animal husbandry. The transition could involve several changes that either occur simultaneously or follow in sequence. Together, this bundle of transitions has presented a key puzzle to archeologists, and debates about *where* and *when* it happened, *why* it happened, *how* it happened, and to what *effect* have defined archeological research for decades.

The implications of the Neolithic Revolution are considered dramatic. Archeologists have taught us how domestication laid the foundation for accumulated wealth and population growth and brought in its wake sedentary societies, the production of surplus, social stratification, centralized political structures, and state formation. Fragments of bones, middens, and broken pots have become the stuff from which we can map our shared origins and piece together the story of how we came to be who we are. Archeology, in other words, offers an indispensable source of knowledge through which at least part of humanity can celebrate, or ponder, its achievements as well as its failures. In this way, it serves as a kind of world-making in which civilization comments on itself.

This master narrative is centered on Europe. Childe's cultural approach to prehistory in the 1920s, which gradually replaced the previous evolutionary models, is a typical example. His book *The Dawn of European Civilization* inspired twentieth-century archeologists, whose work became significant for national projects (Trigger 1996). While much archeological work on domestication focused on material utility, Ian Hodder's *The Domestication of Europe* (1990) took a different view in proposing the possibility that "domestication in the social and symbolic sense occurred prior to domestication in the economic sense" (31).[19] Central to his argument is the opposition between the domus and what he calls the *agrios*, meaning "the wild, or savage." Agriculture thus becomes a "culturing" of the "wild," and the origins of agriculture are connected, via the domus-agrios opposition, to the more general process of social and cultural domestication (86). Hodder's structuralist reinterpretation of archeological evidence stirred some debate in archeology but less so among cultural and social anthropologists, for whom the concept of domestication was hardly a central topic at the time.

Faced with Darwin's thesis that we were once "completely animals,"[20] the domestication narrative offers an idea of progress. On the one hand, it is a story of all of us (cf. human universalism). On the other hand, it offers a tool for differentiating human groups from one another (cf. human multiculturalism) in relation to an imagined, singular trajectory towards a version of "European civilization." This popularized narrative of domestication does not do justice to more recent archeological research Equipped with ever-more sophisticated techniques for reading bones and pots and plant remains (genetics, soil analysis), archeologists have produced more nuanced evidence and hence different stories for thinking about how domestication may have evolved.[21] Nevertheless, the idea of a Neolithic Revolution has had a profound impact on the Euro-American imagination for more than a century, and it has become, to quote Strathern again, "an idea to think other ideas with" (1992, 10).

Domestication in Social and Cultural Anthropology

If, as Darwin suggested, selective pressures could lead to morphological changes in domesticated animals, could similar mechanisms affect humans too? This question inspired Franz Boas ([1911] 1938), who argued against contemporary theories of social evolution and sought instead to explain difference through diffusion and to understand how bodies, animal as well as human, are shaped by their living conditions. Boas's concern was to think human bodies, their environments, and their social practices *together*. The abuse of physical anthropology by the eugenics movement brought an end to such speculations and has led to what Helen Leach referred to in 2003 as the "virtual disappearance of this theory of human domestication from post 1950s anthropological writings" (358). In a seminal article, she argues that humans have coevolved with their animals in a physical or morphological sense, including "gracile features" and a less "robust" body type. The idea that the "unconscious selective pressures" involved in animal domestication also operated in humans as they adopted sedentary lifestyles is controversial,[22] and it has had limited impact on social and cultural anthropologists, whose focus remains by and large on the malleable, sensory, and transformative dimensions of the human body rather than with our genes and bones.

Definitions of domestication in social and cultural anthropology typically highlight the human side of the relation, focusing on animals as property, their utility for humans, or their metaphoric or symbolic meaning (see Leach 1964; Levi-Strauss 1966; Evans-Pritchard 1964). Domestication has also been used to designate metaphorical taming (Goody 1977). Many refer to archeologist Juliet Clutton-Brock, whose definition of domestic animals emphasizes control, captivity, and human profit (1994, 26). Evans-Pritchard's classic work *The Nuer* (1964) is a reminder that anthropological interest in human-animal relations is not new, and his attention to symbiosis and mutuality could be read as an opening towards greater attention to the nonhuman part of the relations of domestication. His take on symbiosis is worth quoting at length:

> It has been remarked that the Nuer might be called parasites of the cow, but it might be said with equal force that the cow is a parasite of the Nuer, whose lives are spent in ensuring its welfare. They build byres, kindle fires, and clean kraals for its comfort; move from villages to camps, from camp to camp, and from camps back to villages, for its health; defy wild beasts for its protection; and fashion ornaments for its adornment. It lives its gentle indolent sluggish life thanks to the Nuer's devotion. In truth the relationship is symbiotic: cattle and men sustain life by their reciprocal services to one another. In this intimate symbiotic relationship men and beasts form a single community of the closest kind. (36)

This aspect of his work was, however, soon forgotten, and anthropology has remained, by and large, a discipline of the human with few openings for the kind of relationality that Evans-Pritchard described.

But there are a few exceptions. For example, anthropologist Tim Ingold took an early interest in animal domestication. In 1984, he defined domestication as the way in which humans bring about "the social incorporation or appropriation of successive generations of animals" (4). Like Russel, he adopted an archeological approach in which domesticated animals are seen as the "objects or vehicles of relations between human individuals and households" (Russel 2002, 291). Thus the key change in animal domestication is located "not in animals' bodies, nor even in human-animal relations, but in the social definition of animals as resource" (291). In this approach, domestication becomes first and foremost a change in human social relationships.

Ingold (2000) has since challenged the temporary dualism in traditional archeological accounts and proposed including practices such as capture as an early precursor of domestication practices. More recently, Rebecca Cassidy and Molly Mullin (2007) have mobilized the term domestication to draw attention to the shifting notion of the wild and situated it in relation to contemporary concerns in social and cultural anthropology about shifting configurations of "nature." In her introduction, Cassidy draws attention to how scholars, mostly outside of anthropology, have begun to emphasize the mutuality of domestication at the expense of notions of ownership, property, and control (Cassidy 2007, 2). One might expect that as the approach to animals seems to be shifting, this would also imply a shifting understanding of what it means to be human—or of the Anthropos of anthropology. This, however, is not necessarily the case.

The Anthropos of contemporary social and cultural anthropology is generally *not* a figure whose bodily features are shaped through diet and domestication practices, as Boas ([1911] 1938) suggested. Rather, it is a figure who speaks, thinks, feels, and acts, one for whom the body is more or less a given: If the body is malleable), it is so through the agency of the human as a thinking and acting subject and not as an outcome of ancestral practices. Hence social and cultural anthropology admits a body, but a *recipient* body, a surface that can be acted upon rather than a site of intergenerational change and genotypical variation. We could say that a Foucauldian, or "disciplined," body or a "decorated body" is admitted as relevant to the Anthropos in social and cultural anthropology, while a "domesticated body" with "gracile features" generally is not.[23]

But where does the delineation of the Anthropos come from? Maurice Godelier offered an important distinction between animals and humans when he insisted that only the latter have *History*, while the former have natural *histories* that are reproductive consequences rather than an effect of intentional activity on their part (Godelier 1986). Biological anthropologists might disagree, but Godelier leaned on Marx, who famously stated

> In creating an objective world by his practical activity, in working up inorganic nature, man proves himself a conscious species being. . . . Admittedly animals also produce . . . but an animal only produces what it immediately needs for itself or its young. It produces one-sidedly, while man produces universally. . . . An animal produces only itself, whilst man reproduces the whole of nature. . . . An animal forms things in accordance with the standard and the need of the species to which it belongs, whilst man knows how to produce in accordance with the standards of other species. ([1844] 1961, 75, 76)[24]

Note how a contrast between man and animal is enacted through a notion of agency, and how "making history" through consciously working on his or her surroundings delineates Anthropos as an object of study. The possibility of mutual benefit through the human-animal encounter and of intentional agency on behalf of other animals tends to be left out of this equation.

If it is this version of the Anthropos that is social and cultural anthropology's object of study, then it is only natural that anthropologists have mapped the world in relation to how human groups work their respective surroundings to nourish themselves. Hence, we teach and learn about people as rainforest hunter-gatherers, swidden horticulturalists, rice paddy agriculturalists, savannah pastoralists, and so on. Subsistence practices have taken on a defining character as what indigenous peoples *are* rather than what they *do*. Because indigenous peoples tend to occupy territory that dominant groups left behind, the ultimate Other of anthropology is often also the ultimate Other of the domestication narrative. Hence the domestication narrative becomes an anthropological ordering device both in defining the human groups that are most interesting to study and in delineating what it is about them that requires scholarly attention. More recently, a heightened interest in human-animal relations casts nonhumans in the role of ultimate Other, but the fundamental hierarchy still remains.

This makes it easy to forget that, rather than having evolved within a particular landscape since time immemorial, many of the people we study have in fact ended up where they are as a result of a strategic move, as they have been pushed out of or fled more densely populated agro-ecological forms of life. James Scott reminds us that even after the emergence of agriculture and early state formation in Europe, "most of the time people moved back and forth between foraging, swiddening, and more intensive agriculture, depending on demographic and political pressures" (2011, 219). Reindeer-herding Sami emerged as a specialization only after Norwegian colonizing efforts made more diverse fishing and herding livelihoods less viable (Ween and Lien 2012). And the Amerindian societies of South America which capture the anthropological imagination as a remnant from Neolithic times are more likely to be previously sedentary cultivators who abandoned agriculture and fixed villages in response to the conquest (Clastres 1974), becoming, as it were, "barbarians by design" (Scott 2011, 218).[25] These are stories that could challenge and significantly modify the domestication narrative and its temporal and spatial orders (as Scott indeed does). More often, however, they remain timeless ethnographic snapshots of people that "modernity left behind," or of ontological alterity. As anthropologists fail to mobilize domestication in their study of the "other" or fail to appreciate the corrective potential of domestication practices as they unfold in the vast and diverse "no-man's land" of people whose lives are neither unaffected by the agrarian complex nor fully entrenched in it, the domestication narrative remains unchallenged too. Hence, it continues to haunt anthropology by its absence, performing a particular order in which sedentary husbandry and single-species cultivation (the agriculture-husbandry nexus) become the norm, while the notion of domestication as a singular phenomenon is modified but not fundamentally challenged. In this way, a temporal order is maintained in which particular societies that have not (yet) made the transition become relics of a timeless past and are cast as opposites to an image of modern man (human) as an agent able to manipulate nature (but see Anderson, Wishart, and Vaté 2013 for a different approach).[26] Similarly, nature remains on the "wild" side of the equation and belonging to a receding past (see also Anderson 2006).

This book is not an argument for a sharper and better definition of domestication. Rather I take its imprecision and mutability as a generative feature of the field itself and mobilize domestication as a traveling companion, along with the salmon, hoping that the juxtaposition of the two may trigger novel understandings of each. Rather than fixed entities, salmon and domestication are both multiple and in the state of becoming. I believe that an intriguing potential for renewal, conceptually as well as materially, may be generated at this particular interface. But there are other traveling companions too. I have also been inspired by scholars who are not known for their work on domestication but who share a curiosity about the dynamics of heterogeneous practice.

BIOSOCIAL BECOMINGS: TOWARDS A MORE-THAN-HUMAN ETHNOGRAPHY

My focus on *becoming* is part of a theoretical shift in anthropology from meaning to practice, also referred to as a turn from representation to performativity. Inspired by authors who explore such possibilities (such as John Law, Anna Tsing, Donna Haraway, Karen Barad, Gísli Pálsson, and Tim Ingold), I see this as less of a "turn" than as a reminder that, as Tim Ingold puts it, "our task is not to take stock of [the world and] its contents but to *follow what is going on,* tracing the multiple trails of becoming, wherever they lead. To trace these paths is to bring anthropology back to life" (2011, 14; emphasis in original).

For Ingold, "anthropology is the study of human becomings as they unfold within the weave of the world" (2011, 9), and as such it involves an interest in how "living beings of all kinds . . . constitute each other's conditions of existence, both for their own and subsequent generations" (8). Hence, "organisms are not compendia of difference but the ever-emergent outcomes of processes of growth" (8). This implies not a radical shift but rather a sharpening of our ethnographic awareness to encompass the ways in which our human existence is constituted by the journeys of our consociates and an invitation to pay attention to what Anna Tsing calls "a more than human sociality" (2013). It also implies a more serious recognition of how, as we are "becoming with" our companion species (Haraway 2008), we are never alone, but are already and always what Ingold and Pálsson prefer to call "biosocial becomings" (2013).

While the idea of biosocial becomings offers itself as an obvious guide to a study of salmon domestication, it also involves some interesting challenges. One of these concerns the ideal of symmetrical approach, one that decenters the anthropocentrism associated with nature-society dualisms and strives for a more equal balance. Kirksey and Helmreich (2010) launched the concept "multispecies ethnography"—often also referred to as "post-humanist approaches"—to promote a symmetrical approach in studies of human-animal relations. I shall argue that to pay equal attention to humans and animals requires a methodological toolkit that human ethnographers are not currently equipped with (see also Candea 2010).

Instead I have tried to cultivate an ethnographic practice that, rather than being "post-human," seeks to expand and explore the idea of what a human ethnographer can be and do as a situated and embodied sentient being. This involves not a radical break with conventional anthropological methods but rather an expansion of the standard ethnographic toolkit through a subtle exploration at the boundaries of notions such as "the social." I acknowledge not only that salmon can be social too (cf. Tsing 2013) but that more-than-human socialities require an ethnography beyond words.[27] I acknowledge that I, as a human ethnographer-interlocutor, may sometimes have the capacity to relate across species barriers by transforming the barrier into some kind of interface via embodied scientific or materially mediated techniques. In these endeavors, I have mostly followed the cues from my human interlocutors on the salmon farms. Other times I have engaged salmon directly as my interlocutors in ways that my human interlocutors normally did not do. In practical terms, this means that I have positioned myself close to what I think of as immediate, physical human-salmon interfaces: by the pens, by the tanks, along rivers, and wherever else salmon live, grow, escape, or die. In these situations I have sometimes made my body (or shadow) available to *their* bodies (or perception), my senses and affective interest attentive to their world and engaged my imagination to create possibilities for what Vinciane Despret calls "embodied communication" (2013, 51; see also chapter 3).

Ethnography has always been about creating possibilities for embodied communication. The novelty of my account is that it extends this curiosity to nonhuman beings—and not to four-legged companion species like dogs (see Haraway 2008) or charismatic endangered animals like meerkats (see Candea 2010) but to mass-produced, industrial, soon-to-be commodities like farmed salmon.

In spite of a flourishing literature on human-animal relations and a heightened interest in their affective dimensions (e.g., Latimer and Miele 2013; Buller 2013), very little attention has been given to farm animals in this regard. Instead there seems to be a general assumption that industrial farming and affective relationality don't go well together.[28] This book challenges that assumption by exploring ethnographically the limits and extent of affect, sentience, and relationality across different contexts and practices. In addition to noticing and mimicking how my human interlocutors relate to salmon (through observation, practical participation, and conversation), I have explored our mutual capacity to respond as well as to act responsibly in the relation (Haraway 2008). I have tried not to dismiss opportunities for corporeal or affective engagement, but to let unconventional snippets of ethnographic practice add another layer to the account whenever it seemed relevant to the questions that I ask.

The bulk of this ethnography approaches salmon as a material, embodied entity, always locally situated and thus emergent in the ethnography through heterogeneous practices. I have indicated that such practices involve both human and nonhuman, both salmon and other worldly entities, and living as well as inert matter.[29] Tracing such practices and the contexts and connections that are mobilized through them has involved a spatiotemporal journey in which I have zoomed in on particular moments, and slowed down in particular places, sites, or sections that appeared accessible, intriguing, or particularly rich ethnographically. Many of these sites offer the reader a glimpse of live salmon "in-the-making." At other sites, practices can only be inferred in hindsight, for instance, from inscriptions through which salmon are rendered visible in the abstract as well as on paper and on screens, through numbers, images, graphs, or descriptions. These include policy documents, research reports, and websites. Such sites enact salmon too, but differently and allow farmed fish to travel far into the future as well as to many places at once.[30]

My most important source of insight is based on participant observation and grounded in the moist humidity of tanks and marine pens rather than the dryness of documents, folders, and binders. The contrast should, however, not be overstated. As I will show, salmon farms are laboratories too, where inscribing and enumerating make up an increasing part of the workday.

FIELDWORKING WITH SALMON

Salmon farming takes place in many parts of the world. Chile, Canada, Scotland, Japan, and Tasmania are all important locations for salmon aquaculture. I chose to locate this study in Norway, a country that is known as the main producer and is also where it all started in the late 1960s and early 1970s.

My first fieldwork on salmon farming, however, took place in Tasmania in 2002.[31] New to the field as well as to the region, I was on a steep learning curve, and it was a few years until I began to reconsider my material in relation to the idea of domestication (cf. Lien 2007a; Cassidy and Mullin 2007). Once I did, it became increasingly clear that there was a lot I didn't know about raising salmon *in practice* and that a more participatory fieldwork experience would be useful. Furthermore, it seemed urgent to return to Norway, where salmon farming began.[32]

Norway is home to the world's largest population of wild Atlantic salmon, and Norwegian authorities have an indisputable global responsibility for the country's salmon river habitats. At the same time, Norwegian authorities serve as catalysts and promoters of an industry that at the current scale of activity is bound to have a significant impact on waterways, both locally and in relation to the global channeling of marine resources. Accommodating simultaneously both a growing aquaculture industry and the world's largest stock of remaining wild Atlantic salmon is a precarious balancing act. This balancing act, and its practical and political implications, shapes the context and the conditions of contemporary salmon farming in Norway. Many of the practices described in subsequent chapters are responses to this dilemma, or mitigating efforts (such as sea-lice treatment). The threats to wild Atlantic salmon are many, and they date back long before salmon aquaculture came into being. However, the intensity of farmed salmon in waterways that wild salmon also depend on raises several problems, of which sea lice and genetic inbreeding are seen as the most important. This particular constellation of wild and farmed salmon makes Norway both "site and witness" in an ongoing battle (Erbs 2011, 5), and Norwegian national authorities both the hero and the villain in the many divergent and contested accounts of salmon realities and future scenarios. With these considerations in mind, I decided to design a collaborative project that would include both the wild and the farmed. The project was called "Newcomers to the Farm: Atlantic Salmon between the Wild and the Industrial." My second fieldwork on salmon farming, which took place in Norway between 2009 and 2012 on and off aquaculture sites in western Norway, was part of this project.[33]

When I decided to do more fieldwork on salmon farming practices, a closer engagement with studies of technology and science (STS) and material semiotics seemed like a good idea. This was not only because farmed salmon appears as a sociomaterial "hybrid," an example of celebrated wildness situated at the technoscientific interface of mass industrial production, hence lending itself to the classic concerns of STS. It was also because farmed salmon emerge through practices that are themselves multiple, heterogeneous—at once material, conceptual, social, human, and nonhuman—and what we may describe as noncoherent (Law 2004). In this endeavor, John Law has been my fieldwork companion and theoretical guide (see also chapter 2 and the preface). Our collaborative fieldwork involved numerous visits in the field as well as an ongoing conversation, manifesting now and then in coauthored papers. Even though I write this book as a single author, it could not have been written without him. Many of the ideas presented here have emerged through intermittent dialogue and at the generative interface of anthropology and STS.[34] Working next to John, I learned to cultivate an attention to heterogeneous practices not only as they are represented (named, valued, sorted, classified, or interpreted) but also as they *unfold in practice*. It was their generative potential that I was after rather than simply their cultural form.

For example, salmon are fed pellets during much of the day and kept in circular pens, which indicates that there are aspects of salmon farming that echo husbandry practices. With this in mind, one can see the sheep grazing on the slopes by the Hardanger fjord and the salmon swimming inside the pens in the middle of it as variations on a familiar trope. Similarly, when English-speaking salmon farmers in Tasmania speak of the act of killing farmed salmon as "harvesting," it indicates that the similarity has found its way into language as well, mobilizing images of cultivated fields as well as grazing animals. Agricultural and husbandry practices seem, in other words, to be folded into the aquaculture assemblage. Based on these and similar observations, one might provisionally conclude that aquaculture is modeled on agriculture and on husbandry, and hence on "domestication"—which in itself triggers new questions and comparisons.

This is an important observation, one that invites further questions regarding domestication. Yet, as an analytical move this comparison with cows and goats seems also a little too quick. This is not only because the popular understanding of domestication is rather simplistic but also because the connection to terrestrial husbandry fails to question the relation between ongoing material practices and discursive domains. More precisely, the comparison diverts analytical attention away from the generative agency of people-and-things to the self-sealing metaphysics of conceptual categories.[35]

In a similar vein, while I obviously engage in discourse (words are among our most important ethnographic tools), I have also approached words with some caution. Aware that there are always things that cannot be told—phenomena that do not map themselves smoothly onto an available and shared linguistic repertoire—I have tried to listen *beyond* the words and to not get too attached to the smoothness of that which readily presents itself in the form of narrative and coherent meaning. Instead, I have tried to move slowly, to pay attention to the not-yet-articulated, and to ask questions about seemingly self-evident things. This implies that while I acknowledge the similarities triggered by the use of domestication as a comparative concept, I also approach this as only one analytic avenue among many others.

There are several reasons why this deliberate and meticulous slowing down was particularly important for me. One has to do with the way in which my approach to salmon as biosocial becoming involves a more-than-human sociality, which necessarily needs to go "beyond words." Another has to do with the way an ethnography of Norwegian salmon farming is also, for me, an attempt to realize what I think of as a "postcolonial anthropology at home." The latter point requires some elaboration.

TOWARDS A POSTCOLONIAL ANTHROPOLOGY AT HOME: PERFORMATIVITY AND MATERIAL SEMIOTICS

It is often assumed that working ethnographically in "modern societies," one can no longer achieve the ideals of holism that characterized conventional anthropological studies of small-scale societies (Bubandt and Otto 2010). Why does it seem so difficult? First, it is important to distinguish the pragmatic delineations of the field from considerations that have to do with ontology. On the practical side, the sheer complexity and the multisitedness of a phenomenon or a "field" often require pragmatic adaptations. This study of salmon farming is, for instance, *not* a study of salmon fisheries or of terrestrial farming, even though these activities were carried out by some of the people I worked with. I chose instead to follow the networks of salmon across several different salmon production/construction sites, prioritizing how salmon-related practices generate fields across spatial locations rather than how communities generate salmon production (see also chapter 2). One could argue that this study is also an instance of "anthropology at home," in the sense that an aspect of familiarity and shared cultural background connect the fieldworker and her interlocutors. This is partly true, but as Gupta and Ferguson (1997) have convincingly argued, globalizing processes and discourse have made anthropology at home less a question of cultural identity than a question of partial coherence across what nearly always involves some kind of cultural difference as well.

While our cultural references and life experiences were rather different, I shared with my salmon interlocutors on the Norwegian west coast a kind of lingua franca, which we could call standard official Norwegian. I also obviously share a range of common references to Norwegian public broadcasting, national events, politics, and so on. But the overlap was partial as well as asymmetrical: their local dialect is difficult to grasp for an untrained ear, and in the beginning I struggled to follow conversations. My Oslo dialect is, on the other hand, the most common version of Norwegian in public media, so it was they who adapted to me rather than the other way around. The difference is perhaps trivial, but it illustrates a choice that we always face in such "nearly home" ethnographies: I could have chosen to highlight difference, to generate the local Hardanger community as my "field," and immerse myself in the specifics of local history and politics. In doing so, and in insisting on the local community as the most relevant context for my analysis of salmon farming, I would at the same time have made cultural difference a salient analytical dimension, and the translation from strange to familiar a convenient tool for telling the story. The strategy is all too familiar in anthropology, and not entirely wrong. But it would not have served my purpose very well. Does cultural difference of a fairly contingent nature come into play in generating farmed salmon? How, or why, is it relevant at all?

My strategy was different. Rather than look for cultural difference, I paid attention to the familiar, assuming that the concepts and assumptions that we all shared and took for granted (and that even John, a nonnative speaker of Norwegian, often shared with us) would be the most significant and important to unpack. Hence, concepts like nature, salmon, industry, and welfare are far more important to the analysis than the vernacular specifics unique to the Hardanger coast.

Michael Scott, who works in Oceania, has argued that in order to achieve holism, we need to situate our ethnographic practice "relative to the deepest level of ontology operative within a cosmological framework" (2007, 19). Leaving aside the problematic nature of the word "holism" and the question of whether such cosmological frameworks even exist outside of anthropological texts,[36] I find his comment on ontology interesting. Situating ethnographic practice relative to the deepest level of ontology is indeed difficult in Oceania. But it is possibly even more difficult in a field in which the "operative" ontology is shared by anthropologist, informant, and their anthropological and public audiences alike. The problem is indeed part of what it means to do anthropology at home. But the difficulty has, I believe, less to do with the field as such than with our epistemological toolkit.

With relatively little emphasis on formal methodological training ("travel far away and allow yourself to be surprised") and with amazement and the "spirit of adventure" as methodological guides,[37] anthropologists are generally well equipped for a particular type of travel—namely, the journey into the unknown. But how do we question that which is taken for granted? How do we unpack, or problematize, those entities and distinctions that seem self-evidently true? How do we disentangle dualisms like nature-society and humans-animals, when such distinctions underpin discourse in the field, because of our shared reliance on words that are already known, familiar, and shared?

The task requires a careful reassembling of our epistemological toolkit. The spirit of adventure is no longer enough. What is needed, instead, is a persistent and uncompromising questioning of the world as it is and what it is made of: in other words, a deep questioning of that which is taken for granted across broad domains. We are moving here from a concern with epistemology to an unsettling of conventional ontology. This is where things tend to get particularly muddled, especially when the common anthropological distance between "home" and "field," as well as "field" and "audience," begin to collapse, as is often the case for those of us who are fieldworking close to home.

Fieldworking in a language that is radically different from the language of analysis offers an obvious advantage. Consider, for instance, how terms like *mana, hau,* and *potlatch* have expanded European anthropologists' imaginaries of what human worlds may consist of and the kinds of worldings that are possible. Through linguistic differences, and doubtless through misunderstandings too, ontological differences presented themselves as key findings, to be elaborated analytically in ethnographic narratives that made our worlds wider, richer, and more nuanced.

Based on the assumption that it is difficult to notice the culturally familiar (cf. "home-blindness"), challenges facing anthropologists at home have conventionally been solved through anthropological comparison: with the help of terms like *mana, hau,* and *potlatch* the familiar could be made into something strange. Amazement was engaged through the back door, so to speak. This has been a useful strategy, but it is hardly sufficient. In order to compare, one needs first an idea of what *is*, an idea of potential units of comparison, or of phenomena that might differ. But what if it is the very emergence of such phenomena that our analysis seeks to capture? Situating ethnographic practice "relative to the deepest level of ontology" implies that such units cannot be known beforehand. They are generated through fieldwork, step by step, at home as well as "abroad." Hence they are *emergent in ongoing field practice,* and as such they resemble outcomes, *explanandum* (what needs to be explained) rather than *explanans* (that which explains). To use a more familiar terminology, whether an emergent phenomenon is engaged as *context* or as the *focus* of study is far from given; rather, it is precisely through such shifts between figure and ground, context and focus, that the analysis gains momentum. So how then do we study how familiar things come into being? Or, borrowing an image from Karen Barad (2003), how do we capture those subtle moments of intra-agency when that which is emergent acquires a particular form?

This is where material semiotics becomes particularly useful. Often referred to as a branch of French and/or British actor-network theory, but with roots in California too, material semiotics offers a rich and well-developed methodological toolkit for ethnographers who stray into fields where amazement cannot be expected to occur. Among its most important features is what I think of as a relentless, untiring rejection of pieces of "evidence" that are mobilized to construct or confirm taken-for-granted realities both in relation to an everyday experience of the world as it unfolds, in Northern Europe for example, and for social science analysis more generally.

John Law's and Annemarie Mol's version of material semiotics is particularly useful for my purposes. With an empirical focus on heterogeneous practices and through ethnographic engagement in so-called modern societies and/or institutional settings, they tread familiar ground. Their work exemplifies the merit of slow methods in studying phenomena we tend to take for granted and is conscious of the ways in which seemingly stabilized entities emerge. Hence, phenomena such as "body," "atherosclerosis," or "aircraft designs" are simultaneously noncoherent *and* constitutive of the real, and in this way they offer a way out of the problem attributed to cultural "home-blindness" (Law 2002; Mol 2002).

Rather than assuming that nature, society, people, market, gender, ethnic groups, and the like actually exist, it explores how such realities come into being through relational practices in dynamic ethnographic settings. Instead of asking what nature "is," the question is how nature is "done." This turn, which is often also described as a turn from representation to performativity (see Abram and Lien 2011; Barad 2003), is fundamental to material semiotics. It is, I suggest, through this performative turn that we can hope to approach what Michael Scott refers to as the "deepest level of ontology" in our own society. With this approach, domestication does not precede salmon analytically, nor did salmon emerge before domestication: the two are generated simultaneously, as potential "contexts" for one another (but not the only ones).

Material semiotics is often associated with the "ontological turn," but the overlap is only partial. While the so-called ontological turn, or what I prefer to call an ontological opening,[38] thrives on radical cultural difference, the version of material semiotics explored by John Law, Annemarie Mol, Donna Haraway, and others has no need for such difference. Rather, it thrives in familiar institutional settings and is premised instead on a practical ethnographic sensibility. The task is to notice the dynamic and emergent constitution of "worlds" through heterogeneous relational practices—practices in which other-than-human agents are endowed with the potential to act. What emerges then is not one world but many, partially overlapping, sometimes coherent and sometimes not. Rather than a question of different metaphysics, or radically different and incommensurable human or natural worlds (which social and natural scientists respectively have claimed privileged and unique access to), what emerges is a multiplicity of ever-emergent human-natural worlds that sometimes rub up against one another, sometimes cause controversy and friction, and sometimes unfold quietly side by side. The multiplicity thus assembled does not claim any privileged access to incommensurable difference, nor does it require ethnographic "amazement." Instead, it works meticulously on what often appears as the familiar and mundane practicalities of the day-to-day business of living and working. Gad, Jensen, and Winthereik (2015, forthcoming) refer to this STS-inspired approach as "practical ontology" in order to distinguish it from the other approaches to ontological difference proposed by Viveiros de Castro, Holbraad, and others.

The contrast is important. Precisely because it neither presupposes nor reproduces a radically different metaphysics (or an a priori cultural or ontological distance between fieldworker and interlocutor, or between anthropological audience and ethnographic field), practical ontology contributes to what I think of as a postcolonial anthropological toolkit. This is because it helps to decenter the concepts that form the foundation of our epistemological anthropological approach—nature, society, person, market, and even domestication—and that seem so difficult to do without. The way in which practical ontology helps is methodological but also ontological: it helps us to notice, ethnographically and through practical and lateral engagement, how such notions, or concepts, are not self-evident but are constantly stabilized and reaffirmed in practice (see also Hastrup 2011). In this way, they also help us to question what appears self-evidently real.

Let us take salmon and nature as an example. A practical ontology approach takes neither of them for granted nor the current coming-together of the two, as in the fairly recent term "wild salmon" (*vill-laks*). The point is not that wild salmon is something completely new but that the boundaries are ever-emergent and unstable. The boundaries established through the conceptual work of terms like "wild salmon" constitute nature as something separate from the human domain: under the water surface, there is an endless variety of creatures that could possibly be included in the category of wild salmon, but scientific practices render one of these particularly visible—namely, a single species: Atlantic salmon. A particular version of that species—namely, the one that is not fin-clipped and shows no signs of having escaped from a pen—is referred to these days as wild Atlantic salmon (see also chapter 7). For the moment, the entity thus assembled appears to hold together fairly well, but only as long as the author renders the work that is involved in achieving this invisible in the analysis.

With a focus on practice, multiplicities emerge. We see that the various practices that constitute "wild" sometimes overlap, but not always. Just as Annemarie

Mol (2002) describes "atherosclerosis" as multiple, emergent through different diagnostic practices, salmon is multiple too (Lien and Law 2011). We are not talking here of interpretative flexibility, nor are we particularly interested in social controversy as such, or in cultural or logical inconsistency. Instead, we work on the assumption that reality can be not one but many. We are not saying it necessarily is, but we are open to this possibility. This approach, which draws heavily on previous work of Law and Mol, and is further developed in relation to salmon together with Law, frames much of the subsequent analysis.

Such an approach is agnostic in relation to some fairly deep-rooted ontological premises in common Western thinking. This is not easy; it pushes us towards the edge of our ontological comfort zones; and to work in this mode is challenging, linguistically, socially, and even emotionally. It's hard to write, hard to imagine, and difficult to argue within an anthropological discourse that assumes that natural science provides privileged access to matters of the real. But I believe that we have to move through explorations at the edges of certainty if anthropology is to be able to transcend its obsession with "the other" and the epistemological constraints that follow from this colonial heritage. It is through such explorations, I suggest, that we can hope to move towards a postcolonial anthropology at home.

CHAPTER PREVIEWS

This book traces different moments, meetings, and calculations in which farmed salmon become visible in aquaculture. The argument is that the turn to marine aquaculture involves a series of transformative relations through which human and animal are shaping one another. Salmon domestication thus involves relational practices in which the outcome is indeed open-ended. Attentive to scalability as well as to the materiality of human-animal encounters, I trace the fragile and contingent relational practices that constitute domesticated salmon and the multiple ways of "becoming salmon" that emerge as a result. I have organized the material to reflect what I think of as multiple becomings, which then in turn make up the titles of most of the chapters.

Chapter 2, "Tracking Salmon," introduces the reader to salmon aquaculture in Hardanger, western Norway, from the perspective of regional politics and scientific concerns as well as the pens and cages of marine production sites. It situates the industry historically in a region where fish have always been the backbone of subsistence practices as well as international trade and includes a brief introduction to Tasmania, which is an important site for salmon farming in the Southern Hemisphere. The chapter describes the emergence of salmon aquaculture in the 1960s and 1970s, the unexpected profitability and growth of the industry since the 1980s, and the emergence of farmed Atlantic salmon as a global commodity. Drawing on ethnographic examples from both Norway and Tasmania, I also describe fieldworking strategies and emphasize how ethnography is an ongoing negotiation of practical opportunities as well as theoretical concerns. The chapter shows that domestication made Atlantic salmon far more mobile than any previous attempt at acclimatization.

Chapter 3, "Becoming Hungry," introduces the salmon domus as home to hundreds of thousands of salmon, as well as a work site for a handful of people. With a focus on material structures as well as on different modes of human-animal exchange, I ask what, specifically, constitutes the salmon domus. Exploring the implications of raising animals underwater, I describe the water surface as a boundary as well as an interface of human-animal engagement and a crucial site for human assessments of the salmon's health and well-being. I follow the farmworkers on the daily rounds of feeding, collecting dead fish, and tending to the well-being of the salmon, dealing with lice, wrasse, mackerel, and other creatures that gather in this multispecies assemblage. As we pay attention to the many practices that assemble the sites where salmon grow, we may also appreciate the efforts involved in simply holding this giant assemblage together. The salmon domus emerges as a "fragile miracle," where an apparently smooth operation conceals the many ways in which much of what we do is experimental, contingent, and very much in the making.

Chapter 4, "Becoming Biomass," shifts the attention from fleshy salmon to figures and calculations and the ways in which salmon are rendered mobile and comparable as biomass. A farmed salmon is a salmon that is putting on weight, but how is this achieved? Tracing salmon through various inscriptions on and off salmon farms, I show how feeding practices render salmon legible and visible, internally as well as externally, and hence amenable to managerial control. Salmon are enacted as hungry, and biomass serves as a boundary object that aligns the fluid world of salmon biology with the volatile world of the global market. Such translations are a major tool for managing salmon aquaculture, for the farmworkers as well as for financial investors.

Chapter 5, "Becoming Scalable," argues that scalability is not an inevitable outcome of capitalist enterprise but a mechanism that must be examined in its own right. I suggest that in order to understand the unprecedented expansion of salmon farming worldwide, we need to look more closely at how scalability is achieved. The chapter details how "speeding up" and "freezing time" are key mechanisms underpinning the current expansion of salmon aquaculture. I draw attention to the feed pellet as a crucial technology for the globally expanding salmon farming project. The feed pellet in turn paved the way for the massive transfer of marine resources from the South Pacific to the North Atlantic. I suggest that marine domestication may be understood as a series of temporal alignments, carefully crafted through textured attachments as exemplified by the feed pellet. As a point of comparison, I invoke canned Pacific salmon and the industrial expansion that canning technology made possible.

Chapter 6, "Becoming Sentient," explores the simultaneous choreographies of caring and killing. Farmed salmon in Europe have recently emerged as sentient beings, subjects of animal welfare regulations. More precisely, they have gone from being simply fish to being animals in a legal sense, with welfare requirements and rights similar to animals like sheep and cows. In this chapter, I explore sentience as a relational property that unfolds through heterogeneous choreographies of care: legal, technical, or the practical routine of feeding and cleaning. Human sensitivity is only one register that may be evoked in the enactment of salmon sentience; the technological apparatus is another. How these are configured and made to enact salmon sentience is always open-ended. I suggest that salmon are becoming sentient because they no longer suffer alone; they suffer *in our care*.

In chapter 7, "Becoming Alien," I follow salmon out of the pens and up the Vosso River, once a famous fishing site for British anglers but now where wild salmon are nearly extinct. Only about nine generations removed from their "wilder" cousins, escaped farmed salmon are now seen as alien species in their rivers of origin. This chapter explores the interface of "farmed" and "wild" in public and scientific discourse as well as ongoing projects to restore the Vosso as a salmon river. I show how "wild salmon" is a fluid category, one that is being constantly reworked, and how new subcategories of salmon proliferate. The chapter traces salmon along unexpected trajectories, to suburban industrial sites and to places where escaped farmed Atlantic salmon are thoroughly out of place, homeless, alien, and unwanted.

In Chapter 8, "Tails," I ask how domestication may still be useful as a guide to living well together, and how it may be thought of differently. Finally, I consider some pressing concerns regarding the current expansion of salmon aquaculture.

2

TRACKING SALMON

The view from the twin-engine airplane is a mosaic of islands and inlets. Glaciers disappear from view as we descend from the rugged western part of the Hardangervidda mountain plateau and approach Sunnhordland, the labyrinth of waterways where melted snow and glacial water meet the salty seawater from the Atlantic Ocean. The steep valleys and glittering surface below is what tourists know as the Hardangerfjord, famous for its natural beauty. We are westbound, on the only daily direct flight from Oslo to Stord, the center of one of the core regions of Norwegian salmon aquaculture.

Across the aisle, a woman in a smart business skirt and jacket is looking at her manuscript, yellow marker in hand, silently rehearsing. The flight takes less than an hour, and soon we step down on the runway, eyes adjusting to the low January sun. Around thirty people queue up by the windswept airport parking lot, waiting for taxis. We seem to be headed in the same direction, the two-day annual regional salmon conference, to listen to plenaries and take part in workshops. The event, which is referred to as the Aqkva conference, gathers nearly three hundred people, including the senior executives of salmon farming enterprises, veterinarians, scientists, bureaucrats, journalists, and politicians, mostly from the region. But there are also a handful of policy makers from Oslo, such as the woman with the manuscript. She will soon be introduced as the state secretary of the Norwegian Ministry of Fisheries and Coastal Affairs; she will give the first plenary presentation.

Tracking salmon sites requires planes and buses, cars and ferries. It calls for attention to timetables, weather forecasts, and maps. It involves driving in the dark on icy, winding roads and enduring carsickness on land and seasickness at sea. The journeys unfold in a tapestry of roads and public transport that connects the many localities where salmon gather under the surface, in marine pens. I have learned to spot such localities from a distance: circular shapes scattered along the fjord, in repetitive patterns—sometimes four, sometimes eight or more. But salmon sites are more than underwater worlds of fish. Salmon are enacted in a multitude of practices, some of which involve salmon-in-the-flesh and some of which don't.

Tracking salmon means weaving together a few of the many different patches of salmon enactments that I have come to think of as construction sites (Latour 2005, 88–89; Law 2002), sites in which salmon are currently in the making. Some are fairly wet and require the use of boats, disinfectants, rubber boots, overalls, and waterproof sheets of paper. These include locally situated hatcheries and grow-out sites, as well as roe-production sites, where the new salmon eggs come into being. Occasionally, they offer an embodied, sensual experience of salmon skin against a rubber glove, a weight in your hand (see Law and Lien 2013), or even roe entering your mouth (see chapter 5). But most of the time, even here, the salmon themselves are literally out of human sight. Other sites are "dry," like offices and computers. These are sites where salmon manifest as numbers, sorted into columns on a spreadsheet, because salmon emerge on paper too. They are enacted in conversations, in graphs, reports, and PowerPoint presentations. Such propositions are powerful epistemic agents that index past events and propose future scenarios (Knorr-Cetina 1999; see also Asdal 2014). Construction sites thus include the many sites and the many moments when salmon are being enumerated, known, inscribed, or otherwise rendered visible, or real, as numbers, graphs, words, or dots on a map, or lively flesh beneath your feet. And then there are sites like the conference at Stord, where a temporary human assemblage shares various renderings of particular salmon realities and where a kind of calibration is crafted out of multiple, but not entirely coherent salmon stories.

Anthropological studies of matters to do with nature or the environment often see these differences as differences of opinion, perception, or interpretation, differences that do not fundamentally preclude the possibility that there is a singular nature out there, amenable to human knowledge practices. A material semiotics approach sees multiplicity instead as the outcome of heterogeneous practices of enacting knowledge. With John Law, I have traced salmon through an ethnographic commitment to heterogeneous practices. I am more interested in how salmon is done, enacted, or allowed to emerge through practices than in what people say or believe. Their aim is often to achieve a closure. My aim is also to resist that very closure, in order to allow other complexities to emerge. Rather than setting up a contrast here, I think of various salmon, as well as the various approaches to salmon, as complementary—not because they together will constitute a coherent whole, but because, as Gregory Bateson put it, two stories are always better than one.

What we will encounter, then, is not one salmon but many. Or, more precisely, we could say that salmon is "more than one but less than many" (Strathern 1991,

36; see also Law 2002, 3). The salmon we encounter are continuously being enacted in heterogeneous practices in which humans take part, and act, but in which they never act alone. Out of these practices, we can trace the emergence of salmon that are only vaguely similar to each other, only partly overlapping, but rarely coherent. Add to this a temporal dimension, and what we study is not only an entity that grows and transforms as it goes through what is commonly referred to as the "salmon life cycle" (see chapter 3), but one that also changes over time. Selective breeding and specific rearing practices imply that a Norwegian farmed salmon in 2010 is not exactly the same as a Norwegian farmed salmon in 2005 or 1995, let alone a Tasmanian or a Chilean farmed salmon. And yet, through such shifts and variations, one soon learns to recognize a familiar theme. The world is not made from scratch every morning. Most practices soon become habit; they become standardized or otherwise solidify into patterns that make salmon farming a fairly recognizable entity, across continents and over time.

FIELDWORK TACTICS: MULTISITED, MORE THAN HUMAN, AND COLLABORATIVE

A young colleague once commented that one of great things about anthropology is that it offers no place to hide—not in the field, not in the text.[1] Even though she might have overlooked how strategies of silencing are intertwined with the history of our discipline, I take comfort in her more cheerful outlook and her commitment to how things ought to be. This book is grounded in ethnographic methods and in the anthropological practice of participant observation. And yet, in some important ways, my practices differ from the classical anthropological practice of fieldwork as a single individual's immersion at a particular site for a very long time. Fieldworking is, for me, a careful crafting of connections between the researcher and his or her chosen field or topic of interest, a tinkering of constraints, affect, affiliation, and affordances on both sides. Often serendipitous, but also the result of meticulous planning, ethnographic practice is always unique because it reflects the particularities of the field and the fieldworker as well as the specific circumstances in which the two come together. Making all these conditions explicit is impossible, but even so, and as a gesture to my young colleague, I would like to elaborate on those that seem important for my analysis. What follows, then, is a brief account of *how* we did fieldwork, and why.

Because farmed salmon are mobile, we were mobile too. As noted earlier, tracking salmon sites involves numerous leaps to sites that are only vaguely connected, and where the trails we might follow are uncertain, broken, and often contingent. Some of our sites appear permanent and ever-present, year after year: they allow us to come and go and make ourselves at home as we gradually get to know the places, the salmon, and their people. These include hatcheries, smolt production sites, and so-called grow-out sites, which consist of square cages or circular pens situated in seawater.

Salmon arrive at grow-out sites only after they have gone through "smoltification." This is a biological process that prepares young Atlantic salmon for the shift from freshwater to saltwater, a shift that is necessary for salmon hatched in rivers to reach the nutrient-rich North Atlantic, where they can grow.[2] Farmed salmon smoltify too. For them, smoltification marks the transition from freshwater tanks to cages or pens in the fjord. Thus, it marks the end of a process of growth and development, which began with the delivery of fertilized eggs to industrial hatcheries.

Other sites manifest only temporarily, such as the annual Aqkva conference at Stord. Here, salmon figures not only as an object of research but as an object of governance and political controversy, for mayors as well as for state secretaries. It is enacted as biomass on farms, as wild salmon populations in the rivers, as escapees, and as revenue—as regional livelihood and as local tax income. It is also enacted as host to sea lice, a parasite that affects all salmon but is assumed to flourish as a result of the concentration of farming operations; hence, salmon are both vector and victim, depending on their positioning inside or outside the aquaculture pens. Salmon's presence at the Aqkva conference speaks to paper trails and research trials, to local entrepreneurial investments and cautious political balancing acts. None of this is simple, and every epistemic agent leads out of the conference room, beyond the present moment, to sites elsewhere, such as, for instance, a research institute in Bergen, a wild salmon trial in the Vosso River, local business investments, policies of the Ministry of the Environment, or the fluctuation of salmon prices on global spot markets.[3]

Most of our fieldwork was situated in western Norway, in the Hardanger region, where a salmon firm that we call "Sjølaks" kindly granted us access to all its activities.[4] Sjølaks is a medium-sized, locally owned company. Like many other firms in the region, it is expansive. In this way it reflects the development towards fewer and larger holdings that characterized Norwegian salmon aquaculture in the 1990s and 2000s. During the time of our fieldwork, it employed more than two hundred local workers and managers and spanned several counties and more than twenty localities, including hatcheries, smolt production sites, and grow-out sites.[5] Our fieldwork with Sjølaks took place over four years (2009–2012) and involved repeated field visits to various sites; we usually stayed one week at a time. We spent our time in overalls and rubber boots, trying to help out, but we were never on the company payroll. Our efforts to take part were voluntary—and, we hoped, a bonus for the company. By participating in day-to-day tasks, we came to know a small number of production sites, as well as the people who worked at them, and became attuned to changes over time.[6]

Each of these sites can generate its own context, explicitly or implicitly, through specific forms of comparison, contrasting, or translation. If scale is imagined as concentric circles, we can picture salmon being enacted through constant traffic inwards and outwards. But scale is not a thing already out there; rather, it is an epistemic device in its own right, lending itself to human efforts to make worlds meaningful, legible, or profitable (Tsing 2010). Seen in this way, there is no resting point, no obvious center, and hence no obvious site from which other contexts can be deduced or made relevant.

Because salmon are not human, our fieldwork was also a more-than-human encounter—a careful crafting and assembling of different spatiotemporal emplacements through which humans and salmon come together. We can imagine each of these emplacements as "knots" in an unfolding multispecies tapestry, in which the weaving is done not only by humans but also by salmon, sea lice, and other entities known and unknown. As we entered each of these sites and slowed down enough to tune into their temporalities, various layers unfolded through our engagement. We were reminded that each knot is endlessly rich in nuance and simultaneously bounded by specific localized horizons. In what follows I refer to such fieldwork sites as "spaces of relative coherence" or "patches" (Tsing 2015). Anna Tsing reminds us that patches can be at any scale, in principle. But in practice and in light of the constraints and openings of fieldwork, the patches that I explore here are those that allow me to embrace and creatively exploit my human capacities as an "ethnographic device." I am a salmon-eating, family-feeding, digesting female organism, two-thirds of the way through her expected life cycle, whose body is ill adapted to underwater worlds but is fairly good at moving aboveground (thanks mostly to transport prosthetics). Compared to salmon, my sense of smell and direction is hopeless, but my other (alphabetical) literacies allow for different kinds of imagination. They offer a sense of other salmon and other people elsewhere and remind me that my horizon, no matter how much I shift my gaze, is always limited, partial, and patchy. For all of these reasons, this ethnography is not a balanced multispecies account but an attempt to include a more-than-human sociality in the story and to recognize that even if humans make arrangements that make a difference, humans are not the only agents that make a difference in the world.

So how then can I—the human ethnographer—know a more-than-human world? I am aware that, in many more ways than I can name, I *cannot know salmon*, so this book is a careful exploration of the few ways that I *can*. I do not swim with salmon, but I feed them, vaccinate them, count them, hold them, and watch a few of them. I do not study sea lice under a microscope, but I take part in practices intended to kill them, and I count them too. I do not perform genetic analyses or laboratory tests, but occasionally I provide the material for them, and I speak with biologists who do. I engage my capacity as a human being in relations where salmon gather or emerge, but where humans always gather too. Hence, I engage in human practices, as humans engage salmon. Tracing salmon becomes a move in several layers: First, I trace people who trace salmon, engaging in their practices.

Second, I trace other people who trace other salmon or trace salmon differently. This is standard ethnographic practice. Third, I weave back and forth, up or down, or sideways, between the different patches that emerge through these engagements. And then, in what we might think of as a fourth, precarious move, I deliberately search for cracks, fault lines, or openings—incoherences that may lead me to see what the people I engage with don't necessarily see, or maybe even, unintentionally, silence or cover up (see chapter 7). This deliberate search makes this book different from a biologist's account. The biologist's aim is usually coherence. Mine is not. But precisely because my aim is not a singular world, it allows other salmon, other beings, other assemblages to enter the scene, and hence another story to be told.

Third, most of this book draws on collaborative fieldwork (see also chapter 1). While I started out in Tasmania doing fieldwork alone, my fieldwork in Norway is mostly done together with John Law. Thanks to generous funding, we were able not only to include both salmon rivers and salmon farms in our larger collaborative project (encompassing both the wild and the domestic) but also to experiment with joint fieldwork in the aquaculture setting. This turned out to be more productive than we had imagined. Whether John and I were working side by side or on different tasks, we found that our observational capacity, as well as our social networking skills, was hugely improved by the mere presence of the other. We spent most days laboring with fish, and most afternoons and evenings opposite each other at the kitchen table, typing up fieldnotes that we wrote separately but then shared and compared. Perhaps the most surprising lesson was how much we both had missed. "I noticed people while John noticed pipes" captures one such productive difference, and thanks to John, I learned a lot about my own limited observational skills when it comes to mechanical devices. This is more than a comment on personality and gender stereotypes. It also reflects the role of language.

While Norwegian is my mother tongue, John's understanding of Norwegian is limited to the most common, everyday phrases. Hence, it seemed like a practical distribution of labor to leave most of the talking to me, while he made notes on material matters, even though in practice almost everyone working on the farms also spoke English, often fluently.[7] But it also reflects our respective disciplinary training (social anthropology vs. science and technology studies), as well as our theoretical leanings. While an anthropological notion of the social rarely permits nonhuman sociality (but see Tsing 2013), material semiotics aims for a symmetrical approach. Slightly overstated, the contrast could be described as follows: Anthropologists trace contexts, guided by how meanings are assembled through words, while STS scholars trace how contexts are assembled through practice. As a result of my collaboration with John, this book also explores the subtle differences, reciprocities, and tensions at the interface of anthropology and STS.

In the remaining sections of this chapter, I introduce a few selected sites where salmon are known and negotiated, as entry points into a field that will become more diverse as subsequent chapters unfold. But first, let us trace the beginnings.

HISTORICAL SALMON

When did it all begin? To trace the cospecies history of humans and salmon along the rugged coastline of what is now known as Norway takes us back to the last ice age. When the ice retreated more than fifteen thousand years ago and shaped the deep ravines that are known today as fjords and river valleys, salmon migrated from the North Atlantic and up the rivers, where they began to spawn. Atlantic salmon are usually anadromous: they are born in freshwater, migrate to the ocean, where they spend most of their lives, and return to freshwater to spawn.[8] As the ice retreated, the coastal area near the edge of the glacier emerged as a habitable zone for humans as well as for mammals, fish, and shellfish. Archeological evidence indicates that there were human populations along this coastline as early as eleven thousand years ago. Hence, salmon and humans were there together from the very beginning, though the nature of their relationship is obviously rather elusive for us today. But as soon as historical records appear, so do salmon, in both texts and images. For example, five-thousand-year-old petroglyphs found in Alta, Finnmark, depict a finned fish that is interpreted as salmon; numerous place-names in Norway are derived from *laks,* the local word for salmon; and legal texts from the Viking era, such as *Gulatingsloven* (operative until 1274), regulate salmon fishing practices and prohibit measures to stop the salmon's migratory journey on the grounds that "God's gift shall move between mountain and shore as she chooses"[9] (Treimo 2007, 28–30; Osland 1990).

Salmon also appears in Norse mythology. Loke, the trickster who is half god and half *Jotne,*[10] appears in one myth in the shape of a salmon. In this way, he could hide both in freshwater and in the sea. Gísli Pálsson (1991) has pointed out how this myth about Loke's transformations elaborates the contrast between land and sea and Loke as a mediating figure, capable of connecting the two worlds (see also Treimo 2007, 34).

Did the Vikings catch and eat salmon? It is hard to imagine why they would not. Other interventions are possible too: A rune inscription on a rock from the eleventh century, found in an inland valley, states that *"Eiliv Elg bar fisk i Raudsjøen"* (Eiliv Elg carried fish to the Red Lake) (Osland 1990, 13). The fish are likely to be salmonids, and the text indicates that fry were moved between freshwater lakes or rivers. If domestication is seen as practices through which humans and nonhumans mutually inhabit each other's lives and worlds or make space for one another, we could see this as an early instance of salmon domestication. Perhaps it was an early instance of river management to improve salmonid stocks! Eight centuries later, enhancement strategies included humans mixing roe and milt, and eggs were hatched in what we may think of as early salmon hatcheries. Amateur naturalists were often involved in what we might consider an early scientific practice. Henrik Treimo (2007) situates such practices in relation to nineteenth-century nation-building and an emergent scientific governance of rivers and fishing resources. Norwegian state authorities funded the first trial hatchery experiments in 1853 in a river near Drammen, with good results (Osland 1990, 13; Chutko 2011, 21). This is an early example of state-funded salmon aquaculture research. Results were recorded, experiences shared, and further experiments followed. In 1870, in southeastern Norway, two "saltwater parks," or enclosed dams, were successfully established for cultivation of salmonids (Osland 1990, 13).[11] The nineteenth century saw a flourishing of hatchery experiments of many different kinds, and salmonids were particularly popular in Germany, France, Scotland, and Scandinavia (Nash 2011). Similar experiments took place in North America, where Pacific salmon were introduced on the east coast, and in Tasmania, where shipments of brown trout and Atlantic salmon from the United Kingdom transformed Tasmanian rivers. Antipodean efforts to acclimatize salmon in the Southern Hemisphere were mostly unsuccessful, but the trout has thrived in Tasmania ever since (Lien 2005).

Even though anadromous salmon were kept in saltwater confinement for cultivation purposes more than 150 years ago, the artificial cultivation of salmon was not very successful until experiments with floating cages began in the late 1950s. In Norway, these experiments are often attributed to the Vik brothers from western Norway, who in 1959 launched a wooden floating cage in which, three years later, forty Atlantic salmon had reached maturity (Nash 2011, 121). These were the first Atlantic salmon to complete an entire life cycle in captivity, and such experiments inspired many others to try out marine cultivation of Atlantic salmon. We see here a tightening of human-salmon relations as a greater part of the salmon life cycle takes place in captivity.

Raising salmonids for sale (rather than for river enhancement or research) may be seen as an early precursor of mass production. During the 1960s, successful trials with rainbow trout at sea led to further trial experiments with Atlantic salmon, but smolt production was a limiting factor until the late 1960s, when Thor Mowinckel established hatcheries dedicated to smolt production near Bergen, and smolts became more readily available (Nash 2011, 123). With the separate production of smolts in freshwater tanks and the use of marine cages for adult salmon, the operational system for salmon aquaculture was essentially in place. By the early 1970s, it seemed clear that commercial marine cultivation of both salmon and trout would be possible (Osland 1990, 13).[12]

In the early 1970s the question was no longer whether commercial cultivation of salmonids could be done, but rather who should do it and how it should be organized and regulated.[13] This, at least, was the question that was raised in Norway, and this is perhaps when politics begin to really matter. From an industrial point of view, Norwegian salmon aquaculture has been spectacularly successful. But industrial success is not inevitable, and it was far from obvious that it would be Norway—and not, say, Alaska, Scotland, Canada, or Washington State—that would spearhead salmon aquaculture and remain a leading producer for nearly fifty years. A thorough consideration of this question is beyond the scope of this book, but a few comments are relevant. Politics matter a great deal, and it is particularly in this realm that the subsequent development of Norwegian salmon farming differed from that in the English-speaking jurisdictions mentioned above.

So, who should benefit from commercial salmon cultivation, and how should the emergent industry be organized and regulated? These were among the questions posed to the so-called Lysø committee in Norway in 1972, and the committee's proposed regulations (NOU [Norwegian Official Report] 1977:39) left an imprint on Norwegian salmon aquaculture for many years to come.[14] The objectives of these regulations were first and foremost to regulate growth and to strengthen local economic livelihoods in remote coastal villages (Osland 1990, 14).[15] These aims were further formalized in the Aquaculture Act of 1981, which reflects the policy objective to "maintain an industrial structure based on small enterprises, an ownership structure based on local ownership, and a widely distributed industry" (Aarseth and Jakobsen 2004, 8).

Pooling income from various sources, land as well as sea, was a common subsistence strategy in coastal Norway, where the term *fiskarbonden* (fisher-farmer) captures a typical way of life. In this context, the idea that aquaculture could provide additional income could be seen as yet another way to make rural livelihoods robust and flexible; it resonated with Norwegian politics at the time. But how did these regulations matter in relation to processes of domestication and human-animal relations, and how could such tight regulations possibly lay the foundation for what would be a profitable global export industry?

Political scientist Dag Magne Berge (2002) has compared the relative success of Norwegian aquaculture to the situation in Scotland, where growth has been much slower. Although both Norway and Scotland had similar cultivation experiments with salmonids in the 1960s, Scottish production has been more limited; in 2010 its total production was around 15 percent of Norway's.[16] Berge argues that while the early commercial aquaculture trials in Norway were characterized by openness and collaboration between local fish farmers, similar experiments in Scotland were guarded as business secrets. In Scotland, ownership and capital were often multinational; Unilever invested heavily, and standardized practices did not always take the need for local adjustments into account (Berge 2002; Magnussøn 2010, 30). A similar structure of corporate and often multinational capital investment has been the backbone of salmon farming in Canada, Chile, Tasmania, and the United States.

While openness and local ownership are important, they are hardly sufficient to account for the unprecedented growth of Norwegian salmon aquaculture. Yet the comparison is relevant, as it reminds us how Norway is, in many ways, the "odd country" from an international perspective. It helps me account for what in Tasmania appeared to me a peculiar obsession with guarding business secrets, a practice that could even prevent local disease remedies from being passed from one salmon farming company to another. It also places in relief our experiences in Hardanger, where, in spite of the presence of multinational companies like Marine Harvest, best practice is still widely shared, informally as well as through formal and mandatory reporting to local authorities. Hence, regulatory measures work together with local and customary social structures to shape the social formations within which salmon domestication finds its form.

Since the 1980s, Norwegian salmon aquaculture has grown far more than the Lysø committee predicted,[17] and external capital investors have shown an interest in the industry. In 1985, another Aquaculture Act was passed; it abandoned the owner-farmer principle, but the restrictions on growth were otherwise maintained (Aarseth and Jakobsen 2004, 8). As a consequence, potential aquaculture investors looked elsewhere for expansion opportunities. This is the background for the growth of salmon aquaculture in Tasmania and also in Chile.[18]

But the story of the emerging salmon industry in Norway is also a story of a unique application of knowledge about selective breeding and the modeling of salmon on cows, or rather on dairy cattle genetics. Agronomist Harald Skjervold, a leading figure in modern livestock breeding in Norway, was instrumental in establishing a national breeding program for what is now known as Norwegian Red Cattle (Norsk Rødt Fe, or NRF), a hybrid of different local breeds that was developed with the sole aim of efficient and economic milk and meat production (high yields, disease resistance, and so on). Skjervold developed an interest in the early commercial production of salmon and established a research station at Sunndalsøra in 1973,[19] where he applied the principles that had been central in the national NRF breeding program to the selective breeding of farmed salmon. Henrik Treimo (2007, 66) describes how salmon roe was collected from forty-one different rivers and then combined through a program of selective breeding with the aim of enhancing a few key parameters that are compatible with commercial aquaculture. In the 1990s, Canadians succeeded in creating a genetically modified salmon that grew more than four times as fast as ordinary salmon. This led to considerable debate in Norway, and the industry and authorities decided not to follow suit (Treimo 2007, 67). As a result, Norwegian enhancement of broodstock is done through selective breeding only,[20] and genetic modification (GMO) is not an option, nor is it something that the salmon industry lobbies for.

The end of the 1980s was a period of further growth due to more efficient feeding,[21] selective breeding, the introduction of vaccines, and massive investment.

Soon, global supply exceeded demand; prices dropped rapidly, and bankruptcies and mergers followed. The crisis around 1990 led to a massive restructuring and deregulation of the entire industry, including a lifting of the earlier restrictions on nonlocal investors in 1991. Since then, there has been a steady concentration of ownership. In 1990, the ten largest firms accounted for 8 percent of the total production of salmon and trout in Norway. In 2001, the share of the ten largest firms had increased to 46 percent (Aarseth and Jakobsen 2004, 9). In 1994, 49 percent of all Norwegian companies held five licenses or fewer; by 2006, the figure had dropped to 22 percent. Similarly, in 1998, only one company held more than fifty licenses. In 2006, 33 percent of all companies did (Kontali Analyse A/S 2007).

Meanwhile, Atlantic salmon aquaculture continued to expand beyond its natural range, in the Southern Hemisphere as well as in the North Pacific. This coincides with United Nations Food and Agriculture Organization (FAO) recommendations for further increase in aquaculture worldwide (FAO 2008), and the expansion has continued ever since. In less than a human generation, or four or five salmon generations, farmed Atlantic salmon have gone from being part of fragmented, uncertain experiments to inhabiting a fast-growing global regime of industrial-scale marine domestication.

Let us now take a look at Tasmania.

ATLANTIC SALMON DOWN UNDER: A TASMANIAN PATCH

The office building is a small room, barely big enough for a desk, some bookshelves, an office chair, and an extra chair in the corner for visitors. Now it serves as my makeshift work space as I leaf through documents, magazines, and reports that I pick out from Jim's shelves and pile on the floor. Jim is from Sydney and works as operation manager for three Atlantic salmon grow-out sites on the southeast coast of Tasmania, an Australian island state. Since the late 1980s, Tasmania has produced farmed Atlantic salmon for the domestic market and some for Asian export as well. The year is 2002, and I have just started fieldwork on salmon aquaculture.

The office is located onshore, a stone's throw from a site called Stephen's Point, which consists of two sets of eight square cages. Each cage assemblage is connected by metal walkways, and each cage is lined with a double layer of nets. The inside net is called a bird net. The outside net is called a seal net; it is a protective measure against Australian fur seals that gather around the salmon farms. But the seal nets are not very effective. This morning, two divers are busy removing so-called morts, dead salmon, which they collect in white plastic bags. So far they have collected 630 from a single cage, one out of three that a seal attacked last night. The seal has been caught and is being held in a 2×1.2 meter cage near the shore, while the workers are busy cleaning up the mess. The netting is still intact. The seal attacked by pushing its body, which is about two meters long, against the cage netting so that the salmon were pressed together in a corner. In this way, the seal gets easy access to fresh salmon and can pick and choose. Usually the seals suck out the liver and leave the rest of the fish to rot, according to the farmhands, for whom this event is nothing out of the ordinary.

"How have you been going with our furry friend?" Jim asks, as a farmhand comes by the office.

The farmhand, who wears Tasmanian leather boots and a broad-brimmed hat, assures him that the seal seems to be doing all right, and that they are waiting for rangers from Parks and Wildlife to come and fetch it. By the end of the day, the seal will have completed a seven-hour journey from the southeast to the northwest corner of the island, where it will be released near the breeding grounds. The farmhand has already spotted the ferry, and the trailer should be here soon.

Australian fur seals are a threatened species and native to Tasmanian shores. Since most adult males are chased away from the breeding grounds (where a single male controls the territory and a large "harem" of females) and gather near the salmon farming region in the southeast, the seal they caught this morning is likely to return. It could be a few weeks or longer before they see it again. But the seal problem is not likely to be solved anytime soon. According to Jim, the number of seals that attack farmed salmon has exploded lately, and they are all males. Last year, an average of five thousand salmon was lost to the company each month between July and December. Eighty seals were caught and transported north, at the cost of about AU$600 each. On the west coast, there are fewer seals, but more black cormorants, another nonhuman salmon predator, which are not always kept out by bird nets.

Salmon are newcomers on Tasmanian shores. Attempts in the late nineteenth century to acclimatize Atlantic salmon in Tasmanian rivers never succeeded (Lien 2005), and it was not until 1985 that Atlantic salmon were once again introduced, this time in an attempt to establish industrial salmon aquaculture. Salmon eggs were sourced from a strain of Canadian broodstock that had been kept in lakes in New South Wales since the 1960s. Nearly everything else was modeled on the emergent Norwegian industry, which by then had turned out to be rather profitable.

My search for Tasmanian "origin stories" brings us back to Harald Skjervold once again. People who took part in the emergent industry tell the story of how Professor Skjervold came to visit around 1984, invited by local Tasmanian authorities to consider the potential for a salmon industry in the island state.[22] According to the story, Skjervold was very encouraging and pointed out that the Tasmanian estuaries were very suitable for salmon farming.[23] Soon after, Norwegian and Tasmanian investors set up a joint venture, and more companies followed. This geographical expansion of salmon aquaculture to the South Pacific coincided with similar investments in Chile and was partly due to the Norwegian policy described above that aimed to secure local ownership in salmon farming and prevent large capital investors from reaping the profits of what was intended to be a source of supplementary income for farmers in coastal communities.

By the 1990s, following a financial crisis in the Norwegian real estate market, the Norwegian investors had pulled out, and Australian owners had taken over. When I arrived in 2002, salmon aquaculture was well established, locally owned, and producing around 15,000 metric tons of Atlantic salmon annually. This made Tasmania a small producer globally but an important supplier on the Australian market.

As I leaf through Jim's trade magazines, I am surprised to find binders with complete collections of the Norwegian magazine *Norsk Fiskeoppdrett* from several years back.

"Do you read Norwegian?" I ask.

"Not really," he says, "but I can understand a little, look at pictures, and I look up some words if I need to."

Jim pulls a Norwegian-English dictionary from his bookshelf and tells me about the five trips he has made to Norway and how important it is for the local industry to keep up with the latest developments. Norwegian practices are state of the art, he says. He adds that when it comes to salmon aquaculture, Tasmania is, after all, a peripheral place.

FROM ANTIPODEAN "PERIPHERY" TO GLOBAL "CENTER"

Salmon aquaculture can be told as a story of globalization, of spatial (geographical) expansion of standardized practice and technology and a pool of practitioners who gain experience in one part of the world and apply it somewhere else. Aquaculture technology, financial capital, genetic material, and fish feed are mobile entities that cross the world in intricate, repeated patterns, making salmon aquaculture what many would call a truly global enterprise. And yet, as the Australian fur seals and native black cormorants remind us, each site is a different human and nonhuman environment, presenting different sets of problems and challenges. Tasmania does not have sea lice, but it does have predatory seals. It does not have ISA (infectious salmon anemia), but it does have amoebic gill disease. A host of other conditions differ too, from water temperatures to legal regulations. Thus salmon aquaculture can also be told as a number of local stories.

But this book is not a story of a food-industrial practice pinned "between the local and the global." Rather, I see global centers and local peripheries as constantly performed and reproduced—the *outcome* rather than the context for aquaculture practices. Jim's Norwegian dictionary resonates with other practices that systematically produce Norway as the global center and Tasmania as the periphery. In Japan, other practices produce other dichotomies, such as that between Japan and the United States (Swanson 2013). Geopolitical hierarchies are also reproduced through the hosting of major international events, such as the biannual aquaculture conference Aquanor in Norway and the biannual Aquasur, in the Southern Hemisphere. Such events make salmon aquaculture a powerful assemblage through which international relations and asymmetries are constantly being made and negotiated. It is precisely *because* salmon can be seen as a standardized figure that comparisons can be made, comparisons through which specific practices are defined as "cutting edge" while others are located "in the past" (Lien 2007b, Swanson 2013, 2015). In this way, and as they engage a linear temporality, the standardizing practices of global salmon farming become a spatiotemporal ordering device that renders some localities ahead and others behind, some central and others peripheral.

Centers and peripheries are relational properties and are constantly made by juxtaposing numbers, as I have just done. But they also emerge through other comparisons. The salmon production of Norway not only outnumbers that of most other nation-states, but it occurs in a country with a population of only about five million people, in which the main export commodity is oil. So seafood is a number-two export commodity, and with aquaculture bypassing wild-caught fish in revenue, the industry is not a negligible source of export income. While Tasmanian salmon aquaculture is not a matter of great concern for federal authorities in Canberra,[24] Norwegian salmon aquaculture is something that national authorities in Oslo cannot afford to ignore.

The staggering growth of salmon farming makes Norway not only the world's leading producer of farmed salmon, but also the second-largest exporter of fish and fishery products, after China (FAO 2008, 48, table 8). More than a million tons of salmon a year are equivalent to a daily supply of 12 million salmon meals a day (FHL 2011). It should come as no surprise, then, that the Research Council of Norway funds applied research under headings such as "HAVBRUK" (Aquaculture),[25] and that news about the salmon industry frequently appears in the business sections of leading national newspapers. Norwegian authorities' interest in and support for research in salmon aquaculture has undoubtedly contributed to its success and also to making Norway a leading player on the global scene. Hence, Norway's position as a center in the world of aquaculture is continuously boosted, reaffirmed, and maintained.

Let us turn now to Stord and to the annual conference for the industry that introduced this chapter.

THE AQKVA CONFERENCE: A TRANSIENT PATCH

On the white screen behind the elevated stage in the large conference hall at Stord Hotel, different salmon are being enacted as graphs and tables. It is January 2012, my third visit to the Aqkva conference, and each year more people have gathered. Men still outnumber women by about five to one, and the gathering provides an opportunity for me to get a sense of salmon farming as a sociological phenomenon, a kind of bird's-eye perspective that is not available from the side of the pen. Each person in the room knows salmon through slightly different practices, so the salmon that they speak about are not the same. The conference is an occasion for such divergence to be expressed. Differences *matter*. Is salmon farming in Hardanger a threat to wild salmon in the region or is it not? Do the recent policy measures to reduce the prevalence of sea lice actually work or do they not? What appear to a scholar with an interest in science studies as multiple salmon are, in this context, urgent matters that need to be settled.

With each new differentiation, there is another potential discussion, another emergent divide in the audience, another salmon reality to defend or to resist. Some speakers leave a lot of loose ends, divergences, or uncertainties, but there are also those who seek to establish a consensus, however fragile or temporary. Occasionally, I sense the emergence of a provisional consensus. This conference is one of many sites in which "realities are conversed into shape" (Nyquist 2013, 51).

Some speakers have a gift for diplomacy, and in her welcoming speech, the mayor of Stord reminds us that although "Hardangerfjord" figures as a single destination for tourists, it is "not one, but many" (*ikkje ein, men mange*). Her reference is vernacular geography, and she names parts of the waterways as they are known locally—Eidfjorden, Sørfjorden, Bømlofjorden, and many more—and insists that different local interests are many too. "Sustainability is what we aim for," she says, and "good dialogues are key."

Following the welcoming address, the state secretary of the Ministry of Fisheries and Coastal Affairs takes the stage. The audience is attentive. Her carefully crafted sentences are expected to reveal the state authorities' policy on further aquaculture expansion, with immediate consequences for local industries and *their* investment policies. She declares that the current government is committed to growth, but that growth must be sustainable. She says that the goal of the authorities (*regjeringen*) is to make Norway the world's strongest seafood nation, and a white paper is under way in order to explain how this will come about. In the meantime the government will continue to finance research, particularly that which addresses areas of heightened concern, such as the genetic changes of wild salmon that result from inbreeding and the recent outbreaks of sea lice in the region. The latter is a grave concern not only because it affects the farmed salmon, but also because the sea lice that inhabit salmon waterways make the smolts' journey to sea more risky and could threaten the sustainability of wild salmon populations. Because the challenges are particularly acute in this region, the authorities have decided to limit the total biomass in Hardanger to 50,000 metric tons until these challenges are solved.

The number triggers relief as well as disappointment in the audience. High salmon prices the last few years have been followed by new investments in the industry. But because each operator is obliged to limit production to a total allowable biomass (operationalized as a defined fraction of the total allowable biomass in the region), the number she presents specifies a limit on growth. Some salmon farmers are relieved, as they feared that the number would be even lower. Others had hoped for more opportunities for further investment and are slightly disappointed. Environmental NGOs welcome the attention to wild salmon but would also have liked to see further restrictions. No one seems surprised by the number though, which appears to be the outcome of a careful balancing act. The state secretary ends her talk by asserting that the acceptance of salmon aquaculture ultimately depends on the industry and pleads: "Make the necessary investments, follow the rules, and use common sense! We will never impose on you everything that is smart, nor will we prohibit everything that is stupid. I work every day to secure a good framework for Norwegian aquaculture, but it is the industry itself that owns the environmental challenges that it currently faces."[26] By dinnertime, the state secretary will have returned to Oslo, while remaining participants mingle in the bar. If this year's conference is anything like the one last year or the year before, the three-course dinner will last several hours, thanks to frequent interruptions by a local stand-up comedian. There will be plenty to drink, and his brief skits will trigger bursts of laughter as well as numerous remarks in various versions of the Sunnhordland vernacular that most participants can locate with confident precision to one of the islands nearby.

At the end of a long day of precarious calibration of policy, best practice, and recent research (including results from my own project, which I, introduced as the "professor from Oslo," have been invited this year to present), the ambience of the evening welcomes laughter and the playing out of alliances other than those emerging from contested salmon realities. I see various relations affirmed and renewed, while I slip into my familiar role of insider/outsider. As the evening comes to an end, we are told that while it is well known that a certain percentage of returning salmon "wander astray" (*feilvandrer*) to a river other than where they "belong,"[27] we are warned not to follow their example but to make sure that we return, each and every one, to the hotel rooms where we belong. This joke is just one of a series of far more explicit jokes with an erotic undertone. My own Oslo dialect marks me as someone who does not belong, and I am also part of the minority as far as gender is concerned. During dinners like this I sometimes I sensed the occasional glance from strangers across the table: "Does she laugh? Is she offended? Does she get it?" After three years of on-and-off fieldwork in the region, I usually do, and I sense how a shared language, like salmon, provides a common ground, as well as subtle boundaries. This feeling of being "other" in an overtly masculine environment was, however, restricted to this particular conference. Although men generally outnumber women at grow-out sites too (where often there are no women at all), there was none of the stereotypically gendered, sexist paraphernalia of masculine work sites on the platforms. This made the salmon production sites in western Norway somewhat different from those I had visited elsewhere, and, according to John, unlike the workplaces that he had known as a young man in the United Kingdom.

. . .

My journey with salmon could be described as a move from an antipodean "periphery" to a global "center." But if you walk along the narrow metal platforms that link the salmon cages in Hardanger, there is not much to remind you that you are at the "center of the farmed salmon universe." With the exception of the seagulls and the constant sound of feed pellets, the place is quiet and stereotypically remote, in the way that tourist brochures promote out-of-the-way places. It is beautiful too. And on the surface it is very similar to other places where salmon grow—in Chile, in Scotland, and in Tasmania. But the stories people tell are different, as their biographies reflect salmon entanglements that are always specific, always situated in a particular time and a particular place in a way that, just as my story here, makes them unique (cf. Haraway 1988).

Let us turn to Hardanger and the island Bømlo.

REBECCA'S PATCH

We have arrived in sunshine, on a narrow gravel road that took us to the water's edge. Rebecca suggested John and I meet her here, by the small pier and the two-story house that serves as office for her salmon farming operation. It is July, and the drive has been breathtakingly beautiful, past lush pastures, sheep and cows, and now and then a picturesque harbor or a small hamlet. We have followed a winding road up across steep hills and down to the coastline; it turns sharply left, then immediately right, as if it cannot make up its mind where it is headed. John misremembers a joke about God, who was so tired after having created the entire Norwegian coastline that he decided to draw the rest of the world's coastlines in simpler strokes.[28] As we park our car, a tall, middle-aged woman in a bright red overall comes to greet us. She welcomes us with a smile and then gestures for us to wait for just a moment while she gives instructions to a much younger woman, a temporary summer worker, who has waited for us to come down before driving an old truck up the steep and narrow gravel road. Avoiding fourth gear is clearly important, but otherwise not much seems to be at stake in this brief exchange about the old vehicle that they both have to wrestle with to get their job done.

We are on one of the western islands in the region. Along with trade, fish has been the backbone of subsistence here since time immemorial. This is where herring has been salted or canned and exported for centuries. This is also where salmon return to reach rivers further east, and until the mid-1970s, when new regulations placed restrictions on coastal salmon fishing, salmon was an important export commodity too. In addition, there is an abundance of fish of many varieties, which makes the region viable in spite of limited farmland. This is also the region where the oil industry triggered a whole new set of opportunities, soon competing with peasant farming and fishing as a new source of livelihood. We are in the region where the social anthropologist John Barnes did fieldwork in 1952 and coined the concept of "network" to analyze social class in a society valuing equality (Barnes 1954). These are some of the things I think about, and my associations point to other books and other stories, but none of these enter our conversations with Rebecca. For now, we let her be our only guide.

Rebecca is the operation manager for this and two other localities nearby. She is one of the few female managers in the region and also among the most experienced. Although she does not own this site, she has been the main person responsible here for nearly twenty years, and she has been in the industry much longer. Right now, she is eager to give us a tour because, as she explains, she needs to check the feeding anyway, so it is good timing.

The salmon are only a stone's throw away, five cages in a row, positioned perpendicular to the sandy shoreline. A metal ramp connects them to land, and there is no need for rubber boots as we walk the forty or fifty meters to the first cage. An automatic feed distributor in the middle of the pen is connected by a pipe to the feed container on shore. We watch it turn around and around, and with each turn, a batch of pellets is sprinkled evenly on the water surface. But Rebecca's attention is with the fish. She wants to know about their appetite, which she can tell by how lively they are just under the surface as they come up to eat. As we walk and talk, she repeatedly insists that the most important thing to remember in this business is to always keep an eye on the fish (*følgja med på fisken*). You don't want to feed too much (no pellets should go past the fish), but you also should not feed too little; in order to achieve this, you've got to watch them very carefully. Incidentally, she adds, she really likes doing this. The only problem, she says, is that current requirements for electronic reporting to the company headquarters tie her to the office. She doesn't mind the paperwork but worries that her attention may sometimes be diverted away from the fish.

The water is clear, transparent, and turquoise blue, shimmering in the midday sun. We let our eyes wander and suddenly spot movement outside the cages. Below our feet, next to the platform, is a gathering of pollock (*småsei/pale*). Rebecca doesn't seem to notice them but confirms that they come around all the time. Then, a bit farther out, a different kind of silvery shadow moves swiftly outside the cages. Now it is a school of mackerel, and we could have easily caught one with a dip net if we had tried. The salmon are kept in place by the netting, but they are clearly not alone in this underwater world.

Rebecca explains that the first four cages we pass, which are 24 × 24 meters each, hold about fifty thousand fish apiece. The fifth is bigger, 35 × 35 meters, and holds eighty thousand. Sizes have gone up gradually since she began: "I used to say," she tells us, "that I would *never* have cages beyond 24 × 24 meters, but look—now that is just what I have!"

When this locality first started, the cages were closer to the shore, where the water is shallow. They were also much smaller then and not as deep. Rebecca recalls that in the early 1960s a former owner started experimenting with growing sea trout in the fjord, in hexagonal pens. The pens were made of wood, measured 6 meters on each side, and held four thousand fish each. The same man had opened a fish landing facility in 1956 and later a grocery store, which is where Rebecca first started working for him. She explains: "In the beginning, they made the feed from fish scraps from the fish landing facility. They made the feed themselves, in a big grinder, and mixed in shrimp to get the right color. That was easy, because the fish landing facility bought fresh shrimp from local fishermen, and there was a shrimp peeling facility here too, so there were plenty of shrimp shells available." Sea lice was a problem even then, and Rebecca recalls how they used onion and garlic as natural remedies: "They cut onion in small pieces and packed them in a bag that they hung inside the pen. They thought that the smell would scare the lice away. Later, they used garlic too, and I remember that they used to order huge amounts of onion powder from the local store, which they mixed with the feed."[29]

"Did it work?" I asked.

"Probably not, but who knows. They were just experimenting, they didn't really know."[30]

Rebecca's career as a fish farmer did not start until 1987, after her children had started school. She was trained by the first owner, and when he died some years later, Rebecca ran the operation, first together with another woman and then by herself. At first, there was only one locality, but in the years that followed (1990 and 2002), two other localities were established, and now she is in charge of all three and has several people working for her.

Later, over coffee, we learn about the work she does and also about ways of preparing salmon (barbecued) and cod, which her husband catches daily. She makes *fiskekaker* out of haddock and *lubbesild* out of herring or mackerel. We turn to the farming and learn about numbers, diseases, the various methods being tried out, and the endless reporting, as well as about what she sees as the most important key to success: "What really matters here is that you do not stress the fish (*ikkje stresse fisken*). Sometimes reporting procedures require that we handle the fish more often. If you can avoid handling the fish, that is good. It is also important to make sure they have enough space. Otherwise, paying close attention and ensuring peace and quiet (*ro og fred*) is key.³¹ If that is taken care of, then they tend to be healthy."

SITUATING NORWAY

I have chosen a sunny patch to close this chapter: Rebecca's patch. Or rather, the patch she chose to share with us on this bright summer day. When we met her, things were going well, but it seems that things have been going well for her for quite some time. Others describe her as a woman who knows what she is doing, and they admire her for that. We become who we are through participation in the heterogeneous landscapes where our lives unfold.

I chose Rebecca's patch because it gives us a glimpse of the early days of salmon aquaculture and allows us to consider how salmon and people like Rebecca have shaped one another in this region. Salmon aquaculture on the Norwegian west coast emerged gradually in the 1970s, as in this case, from small-scale experiments by local entrepreneurs. It offered another way to make a living from fish but was not a complete break with the past. As it turned out, it also provided a welcome additional income for smallholders who struggled to make ends meet, and became a reliable livelihood in the coastal regions of Norway, where salaried jobs are often few and far between. Last but not least: although it attracted more men than women, the salmon industry provided opportunities for young women like Rebecca, who, like so many women in the 1970s, were not content to be stay-at-home moms, but needed a paid job as well.³²

I have indicated how the emergent industry was tightly regulated but at the same time strongly supported by state authorities, who take an active role in regional industrial development. Add to this an expanding and well-funded public sector, a fairly efficient governance and state bureaucracy, and exceptionally high levels of trust in the state,³³ and we have a sense of the structural context in which salmon aquaculture has unfolded. On the salmon farms, the presence of Norwegian state and local authorities is experienced in a number of ways, such as in detailed legal regulations, taxation, regular audits, frequent mandatory reporting of a number of different kinds (see chapter 4), and frequent on-site visits by food safety authorities (see chapter 3).³⁴

Many scholars have pointed to socioeconomic equality and equal gender opportunities as key characteristics of the Scandinavian states, so much so that ironic titles like "The Almost Nearly Perfect People: The Truth About the Nordic Miracle" come as a relief. Even though equality is never achieved,³⁵ it remains a strong ideal and an important ethos in Norway. This takes many different expressions, including what anthropologist Marianne Gullestad (1992) has described as "equality as sameness," or the idea that egalitarian values are best realized among people who are equal in the sense of being similar, or "the same." Gullestad based her insight on her own extensive fieldwork as well as on the work of John Barnes, who did fieldwork at Bremnes in Sunnhordland half a century earlier. Although Gullestad's claim is contested (see Lien, Lidén, and Vike 2001), it captures an ethos of everyday life that is very much present in the salmon farming industry. Hence, it is useful to keep in mind that the encounters between salmon and their people often take place within a work environment in which responsibility is highly delegated, where the operations manager is as likely to engage in practical and manual tasks as the farmhands, and where autonomy and the ability to evaluate the situation and make a quick decision are strongly valued as characteristics of a "good employee." This resonates with a Nordic tradition for strong labor unions and worker participation.

. . .

When we met Rebecca, she appeared competent, calm, and content. She has seen generations of farmed salmon pass through her pens (and figuratively through her hands) and the industry develop from the early experiments of her late mentor to the current highly regulated and highly profitable export industry. The leap is enormous if we look at the numbers, but from her particular perspective, it appears less dramatic: A few more meters have been added to the metal walkway between the shore and the first pen; the pens have grown bigger and can hold more fish; more paperwork is required; and feeding is a bit different. But otherwise the basic operations are much the same. And this is probably why she masters it all so well.

She doesn't let the visitors disrupt the rhythm of the day, but lets the schedule of the salmon feeding guide her own and gives them attention when she knows that they need it. Small and large challenges and catastrophes come and go, but we caught Rebecca in a quiet moment, when the machines were doing what they were supposed to, and the salmon were doing the work that they are so good at: eating pellets and putting on weight.

What we witness here is an instance of mutual relating across species barriers: The salmon have become domesticated in her care, just as they have "domesticated" her. The assemblage includes a local summer employee, an old truck, a feed hopper, a woman, a set of cages, salmon penned up, and a number of uninvited species who see the opportunity to feed either on salmon (lice) or on their surplus feed (mackerel, pollock). The outcome is always uncertain, but for now and for the foreseeable future, it seems to be holding together fairly well. Rebecca's patch offers us a glimpse into particular human ways of knowing salmon. Other patches are enacting different salmon, different people, and different stories about what salmon aquaculture is about. This book is a journey through a few different patches, and different becomings along the mosaic of islands and inlets in the Hardanger region, interspersed occasionally with leaps to patches in other parts of the world. Let us turn next to the Salmon domus.

3

BECOMING HUNGRY

Introducing the Salmon Domus

It looks almost like an ordinary house, floating at sea: pitched roof, gray walls, an entrance, and a small window upstairs under the ridge. Until you notice the adjacent low structure of nets and railings stretched out along the water surface to the right, you might think this is a way to beat the current Norwegian property boom: a houseboat situated in the middle of the fjord. It would be an attractive site for a vacation, sun shimmering on the nervous water surface and a hazy view of the glacier in the distance. It is indeed a beautiful midsummer morning, but we are not on vacation, and this is not an ordinary house but a salmon grow-out locality called Vidarøy, "home" to more than 600,000 salmon and my workplace for the next couple of weeks. (This is also where John Law and I will do joint fieldwork off and on for the next four years, but on this first summer visit I arrive alone.)

We are four people on the boat: Fredrik, the local operation manager, who is in his mid-twenties, and two other employees—Aanund, a young man just out of high school, and Karl, a man in his late fifties who has moved to Norway from Poland with his family in order to work and then retire. Karl had visited the region many times and was captivated by the beautiful landscape and the quiet lifestyle. Five years ago he made the move. There is plenty of work in the aquaculture industry, and he makes a decent living now. Aanund could have been my son. Soon I will be his apprentice. He is only nineteen, but it is already his fifth season as a summer worker, and he seems to know all he needs to know to fill in for the permanent staff, who are now beginning their four-week summer vacation. And then there is me: a professor from Oslo, familiar with salmon farming from Tasmania, but with little experience in doing this kind of manual work.

The firm has granted us access to all their localities. We have studied them on the map on the firm's website and in the print version on the wall inside the office building onshore. A red pin marks each sea locality on the map, a blue pin each hatchery. The pins are many and scattered across several counties and municipalities along the fjords and islands of this coastal region, where fishing has always been the backbone of the economy. We have visited some of them but decided to spend time at Vidarøy, partly because it can be reached conveniently by boat from a village where we can stay for free, but also because it consists of ten cages (and one pen) connected by metal walkways and a platform, unlike the circular pens that are often seen in this region, which can be reached one by one only by boat. The platform walkway allows easy access to where the salmon are and offers a place to rest for professors who are bound to be idle, or simply in the way, part of the time.

The salmon number more than 600,000 and are penned inside the cages. We humans number four. Together we make up the smallest functional unit of what we might provisionally think of as a salmon domus. As I will soon learn, there is more to the domus than meets the eye, and there are creatures inside that are neither counted nor named. The boundaries are fluid, quite literally, but in a figurative sense too, they are unstable and shifting. The salmon will be ready for slaughter this week and next; each one weighs nearly 5 kilos. They consume about 3,000 kilos of feed pellets per day. The pellets are delivered by feed boats every other week or so, and the plastic bags of 200 kilos each fill the storage room from floor to ceiling. Without the pellets, there would be no salmon.

The lifeline of the salmon domus reaches far beyond this picturesque seascape, and beyond the Hardanger region too. Pellets are partly made from smaller fish, some of which come from as far away as the South Pacific. An infrastructure of global trade in marine resources, another of human labor, sourced mostly from the community and the nearest islands, as well as infrastructures of electrical grids, legal regulations, heavy machinery, and tools are all folded into the salmon domus in complicated ways. We shall return to the question of feed in chapters 4 and 5. For now, it is enough to notice that the salmon's house at sea *looks* a lot like a home for humans. But that is only from a distance.

Our boat is going fast and hits each wave with a hard, rhythmic *thump*. Soon Fredrik slows the engine to a lower pitch, and we catch the waves more gently; then he maneuvers the boat alongside the platform and tightens the rope that holds the boat in place. I notice that the flat structure to the right is actually enormous, reaching out in the distance, with a complicated structure of nets and railings more than 3 meters above the water surface. I hear something flapping against the water surface, an unfamiliar sound that I will soon come to recognize as the sound of salmon in the morning.

We climb from the boat up a small ladder onto a narrow walkway and place our boots in a shallow bath of disinfectant before we turn a corner and pass through

FIGURE 1. Vidarøy from the boat (photo by John Law)

the door to the offices inside. From now on, things happen very quickly, a silent choreography of three people moving briskly and confidently towards different tasks. Aanund goes outside while Karl disappears down into the basement, and soon we hear the roar of an engine: the generator is turned on. Fredrik boots the two computers, logs in, and opens a window with an image that I soon learn to recognize as feed silos. Now they appear as purple rectangles on the computer screen, with a number on each. Is there enough feed left in each of the silos to get the feeding system started? He decides there is, and with several maneuvers on the keyboard while watching intently the numbers that appear on the second screen, he shows me that there are four different feed lines, which branch off to the different cages and pen that are numbered one through eleven. Aanund enters, opens the window, and writes down a few numbers on a sheet of paper. "It is even warmer today," he says, "19°C at 3 meters of depth, up 2° since Sunday. Warm enough for a swim!" But this is a recurrent concern during what has been a week of unusually hot and dry summer weather in the region. Salmon prefer cooler water temperatures.

Soon we hear a shattering sound, like a low pressure drill turned on somewhere in the distance. Fredrik nods toward the structure outside: It is the sound of pellets shooting through a bundle of black plastic and metal feed pipes, which are hooked onto the upper structure of the platform and branch off in the distance to reach the cages further out. Feeding has begun. And while the smell of brewing coffee fills the room, Karl comes up; having checked the feed silos downstairs, he reports that there is very little left in number 2 and it will soon be empty. Fredrik checks the screen again, and together they decide that the numbers must be wrong. According to the computer, the feed silo should be half full! Did anyone forget to enter the numbers for yesterday? They look a little worried, but not for long. Fredrik murmurs that this is a hassle, but never mind; he can calibrate the system later.[1] He pulls back his chair, and we all gather around the large, rustic pine table, where Aanund has placed four cups and teabags and has begun to pour the coffee for Fredrik and himself. The morning sun brightens up the room, and I squint and pull the curtains a little. It is 7:30 a.m., and the first break begins.

Karl stretches, sighs, and complains about the heat. "This is not what I moved to Norway for," he exclaims as he reaches for his cup of tea. The other two smile at him, shake their heads, and couldn't agree less. A local summer festival is scheduled this weekend, with concerts and a two-day rowing race (row around your island in locally made wooden boats), and lots of visitors are expected. Fredrik and Aanund need to buy entry tickets for the festival; they can do it online. The conversation flows easily. I learn that there will be slaughtering sometime soon: several of the pens are ready, and they are waiting further instructions from the main office about which ones and when. And then there is disease. Fredrik worries about pen 11. PD (Pancreatic Disease) is on the rise this season, and it affects the fishes' appetite and well-being. Fredrik thinks pen 11 may be affected and has called for the local vet. He hopes she will be coming out here in the afternoon. Can someone pick her up? Aanund volunteers. Karl tells about a fishing trip over the weekend, with a brand-new rod, a gift from his brother. He caught cod in the fjord, and caught his finger on the hook too, and it hurt! They comment that it must hurt the fish as well.

Aanund says that yesterday he caught a nice big salmon in a river not so far from where we are. Nobody seems to wonder what kind of salmon he caught, but my mind is pondering current debates in the media about wild salmon and escapees, so I ask him anyway: Was it wild or was it an escaped farmed salmon? Aanund says it was really hard to tell. The fins were sharp and pointed, which indicates that it wasn't raised in a pen,[2] but he thought he could spot a small mark under the belly from vaccination, which could indicate that the fish had escaped as a smolt and then migrated to the river nearby. So he thinks that it was a salmon bred by the industry that escaped at an early age.

After about half an hour, Karl picks up his cup, walks over to the kitchen, and places it in the dishwasher under the kitchen counter; the rest of us follow suit. It is time to move on. After a brief discussion, Fredrik and Karl decide that the professor can probably be most useful with the *daufisk,* which is Norwegian vernacular meaning "dead fish"; in English the term often used is "morts." Could Aanund please show me how?

DEAD FISH IN THE MORNING

We put on rubber boots and light nylon overalls and walk outside. The sun is high in the sky already. Aanund fetches a wheelbarrow and a bucket and hands me a pocketknife and a pair of blue rubber gloves, and we start towards the end of the platform. From the office he has brought a sheet of paper attached to a clipboard, with a pencil tied on with a string. He explains that he has already been down in the basement and turned on the air compressor. Now we are walking down the middle of the main steel walkway along the long, rectangular platform that reaches out from the building. As we walk, we pass the square salmon cages, 25 × 25 meters each, five on each side. We head towards the end of the walkway to pen 11, a circular plastic structure a bit bigger than the rest of the pens. It is better to start far away, he says, and then work your way gradually back. I soon get the point: the wheelbarrow will fill up with dead fish and eventually be emptied into a large cement container right next to the office. The platform is 125 meters long. The shorter the distance you need to push a full wheelbarrow, the better.

Removing dead fish is one of the first responsibilities of untrained farmworkers, and we were no exception. This is because the farmhands who look after this "city of fish" are also its undertakers—dead fish must to be removed from the living, so death becomes very visible for most humans engaged in the salmon farming enterprise. Finding a few dead fish every morning does not cause any particular concern. It is simply part of the morning rounds. On an ordinary day we remove between one and ten from a cage that contains around fifty thousand fish. When the numbers are small, removing them is a matter of hygiene and confirmation that things are going well. It is a practice of care and a way of paying attention (see also chapter 6). But how is it done?

> One end of a large plastic tube is hanging on the side of the pen; the rest disappears down into the water. Aanund picks it up, hooks it onto a square plastic container, called "Air-lift," tightens it with a rope, and turns a handle. After a few seconds a gush of water comes out, and we laugh as we get sprayed. Then fish suddenly come shooting out, large salmon, mostly dead or nearly so, and a few little ones that aren't salmon at all. They are colorful, unfamiliar, and about the size of the palm of my hand. *"Leppefisk,"* Aanund explains. They are placed in the pens to eat salmon lice, which are a big problem now. The container fills up quickly. Aanund pulls up the sleeves of his overall and reaches down among the dead salmon for the *leppefisk,* which seem to be very much alive. He invites me to help

him, and together we search them out with our hands and throw the live ones back in the pen, leaving the dead ones down there with the salmon. Then we empty the container by picking up salmon by the tail, one by one, and throwing them into the wheelbarrow. It fills up quickly. Shouldn't we count them? "I have counted them already," says Aanund, and tells me the number: 58.

He asks me to fill in the numbers. The sheet, which is made of waterproof paper, has columns and rows, and a few numbers inscribed in the left-hand columns. I notice that there are dates from left to right, and someone has written 30/6: today is the last day of June. Underneath, there is another set of numbers which I recognize as the water temperatures referred to in the morning. I pick up the pencil and write "58" in today's column on the bottom row, marked "11" for the pen number. Soon the wheelbarrow is filled to the brim, and Aanund guides it carefully back along the metal walkway and up the ramp to the cement container, where the dead salmon will dissolve in formic acid. Then he comes back for more.

Fifty-eight is quite a lot, he says, even in a pen that holds 180,000 fish. More than usual, and yet another indication that they may be suffering from PD.

The rest of the round is done fairly quickly, and the numbers are much lower: 5, 6, or 0 are normal counts, I am told, for cages that hold about fifty thousand healthy salmon. But what about the *leppefisk?*

This first year of fieldwork, in 2009, we didn't even count them. A systematic trial of introducing *leppefisk* as a measure against sea lice had recently begun, and about five thousand of them had been delivered to Vidarøy during the previous weeks.[3] Sea lice, a small parasite that was always present in these fjords, had recently become quite prolific on the salmon farms. With the sharp increase in salmon production and a similar increase in the density of production localities in the fjord, there were plenty of salmon for sea lice to thrive on. Sea lice are a problem for farmed fish but even more so for the salmon that hatch in the rivers and pass the salmon farms as smolts on their way to the North Atlantic. In the late 2000s, marine biologists had warned about the high density of sea lice in Norwegian fjords, especially in regions with lots of salmon farming, such as Hardanger. A high number of sea lice in the fjord is bad news; it makes the young salmon's journey to sea more stressful and dangerous and thus poses a threat to the local salmon stock since the young salmon get infested by the parasites on their way to the ocean.[4] As a result, new regimes of more frequent and intensive treatment were being tried out. Sea lice were normally treated with different medications or antiparasite remedies, but each treatment required a major investment of labor, and some of the chemicals posed a risk to workers' health and safety or to the environment. Hence, *leppefisk,* a local "sea-lice eater," seemed worth a try.

During the summers of 2009 and 2010, we saw enterprising teenagers out in their family boats, catching different varieties of *berggylt,* or Ballan wrasse. They took their catch to the salmon farming localities on the fjord and sold it by the bucket, for a small profit.[5] Nobody knew yet whether this would work or whether the wrasse could be kept in cages with salmon. A few years later, in 2012, it was generally agreed that wrasse were a cheap and sustainable way of keeping sea-lice numbers down. They had become part of the salmon domus inventory, counted and accounted for, and a new column had been added to the "*daufisk* sheets" so that the number of dead wrasse could be written down as well.

But on this early midsummer morning in 2009, we simply throw them back in the pen, or into the wheelbarrow, without any further ado. But why do they come out of the *daufisk* pipe when they are alive and healthy? Aanund explains that they probably like to gather near the bottom, around the cylinder that is constructed so that the dead salmon fall down to the center. The plastic tube works by air pressure and sucks up the *daufisk* with such force that the much smaller wrasse are sucked up as well. It is clear that the mechanism constructed for the *daufisk* removal predates the emergence of wrasse as a salmon companion in the pens.

On the platform, Karl is now busy cutting up a pile of black plastic bags, the kind used for waste disposal. He cuts them into long strips, about 10 centimeters wide, and ties them together in bundles of thirty. This is for the wrasse, he explains. This is their cover (*skjul*); they like to hide in seaweed, and this is their replacement, a way to make them feel more at home and less exposed among the predatory salmon.

As we finish the daily *daufisk* round, I learn more about Aanund's passions for fishing, for salmon, and for the region where he lives. There has been a moratorium on salmon fishing in the fjords for years now.[6] But in spite of that, Aanund goes out net-fishing for salmon in the fjord with his uncle, who does work for an environmental research institution. Their so-called research catch produces samples of salmon that can later be analyzed in order to determine many things, such as the prevalence of sea lice, the proportion of escapees, or their genetic origin. But Aanund's knowledge of salmon seems to be linked most of all to his passion for fishing. He tells me that he has been fishing as long as he can remember and knows all the good spots in nearby rivers. In summer, he goes out quite often, sometimes with friends. His dad used to catch salmon in the river that supplies water to the smolt production site onshore, but that was way back, before salmon farming. Now there are no more salmon left in the river, only trout. Aanund thinks it is because out here, on islands near the ocean, the rivers are short, and the spawning grounds are therefore fairly close to the river mouth. So escapees can easily make it up to there, and the wild salmon will be more affected than in rivers further east or inland, where the journey to spawning grounds is much longer and more strenuous.

Salmon farms have been here as long as Aanund can remember, and they are part of his life too. He has seen the installations expand, and while he worries

FIGURE 2. Wrasse cover (photo by John Law)

about the threat they pose to the river salmon stocks, he also thinks it has been kind of neat to see the changes and how they provide work for lots of people now.

Back in the office, Aanund shows me how to enter the *daufisk* numbers into the computer. The data are then sent to the head office, a large, two-story building onshore, where about a dozen company employees staff the offices on a daily basis. This is where major decisions are being made—about slaughter, about sales, about feeding, about delousing, about who goes to work where and when, and about machinery that needs to be moved from one locality to another. As the center of gravity, or the center of calculation,[7] this is also where the owner and directors have their offices and where they gather the operation managers for meetings every Monday morning to coordinate the tasks for the following week. This is where calculations are made that produce or make visible the firm's economic surplus, its profits, or so-called *driftsoverskudd*, and where future strategies and investments are discussed. It is where visitors are greeted and strategic policies nailed down and discussed, and where municipal officers sometimes come to visit (for details, see chapter 4).

But on this warm midmorning on the last day of June, I don't know much about this yet. The shore is hazy in the distance, and all I have learned so far is that numbers are important, that dead fish count, and that the numbers will be passed on elsewhere as soon as they are typed into the electronic forms that are displayed on the computer screen in this office, in the middle of the fjord, where we are now approaching lunchtime. By 11:30, the table will be set again, with fresh bread, cheeses, smoked mackerel, smoked salmon, pickled herring, ham, and salami. A typical Norwegian cold lunch, buffet style, is served every day, paid for by a small deduction from the monthly salary of all employees and a generous subsidy by the firm.

THE SALMON DOMUS

So what *is* this site of activity that we can provisionally think of as the home for farmed salmon, or the salmon domus? How can we describe this biosocial formation, and how does it hold together? I will return to these and related questions in the following chapters, as different practices are highlighted. For now, let us begin by considering a few basic dimensions related to the material structure, such as what it's made of and where it begins.

One of the most important features of this arrangement is the least visible. It is neither the moorings, nor the pens, nor the salmon, but their medium: Salmon spend their life in water. We often say that they are "underwater" or "beneath the water surface," but that is of course a human perspective and reflects that our preferred medium is air.[8] For salmon, the world is three-dimensional. The size of their cages is not 25 × 25 square meters as we tend to say, but rather 25 × 25 × 33 cubic meters, since vertical movements are as important as horizontal: in this world pellets appear in abundance up above and gradually sink towards the bottom. This world is bounded by netting, but not completely. Through the netting flows a constant current. It is quite likely that the flow of water also offers signals to salmon in the form of smell, but this is hard for us to know. What we do know is that the water temperature drops as they move down and that they can adjust their position to accommodate changes in weather and season. They hardly ever see us as human bodies. When they do come really close to us—which for most salmon happens only once or twice in a lifetime—they are nearly always out of water, flapping in a dip net, held firmly by a hand, and always gasping for oxygen, highly uncomfortable or, sometimes, anesthetized.

Standing on the platform, looking down into the pen, I sometimes watch the swift movements of their bodies as they catch the pellets and the simmering of the surface as they jump. Some days a camera is dropped down to the bottom of the cage, and we can watch them on a computer monitor: the black-and-white low-resolution real-time image of salmon swimming by, almost as if we were right there with them. But even so, it is fair to say that most fish on a salmon farm are nearly always out of human sight. Cameras are seldom in use, and when they are, it is usually to solve a problem—perhaps a tube that has been stuck or a device in need of repair—and not to watch fish swimming by. And even if a fair amount of time is actually spent watching fish feed (*sjekke foringa*; see the section "The Water Surface as Interface"), the fish we see from up above are always only a tiny fraction of the fish that are actually down there. We see twenty, or thirty, or maybe fifty, while the number of fish in each pen is at least fifty thousand.

So this is the first thing to notice about salmon domestication. We prefer different mediums. We are nearly always *apart,* separated by the water surface. The material arrangement that makes up the locality at sea can be seen as an interface accommodating this very basic separation. Perhaps we could say that the salmon domus is a device that both facilitates and mediates contact across the spatial distinction between those that need water to breathe and those whose oxygen metabolism requires air.

Many things follow from the way we operate in different mediums: We need air pressure to suck the dead salmon up from the bottom because we cannot move down there and get them. We need cameras to see what is going on. We need platforms, railings, and walkways made of solid steel grilles to keep us safely above water but close enough to the water surface that we can lean over the railings with a dip net and catch a fish. We do all this to compensate for being a species that has lungs, not gills, and that moves most efficiently on two-dimensional planes (give or take a ladder here and there) and are clumsy participants in the three-dimensional water world of salmon. So we built our world around theirs. In a spot with ample depth and currents, we created a structure for *us:* a house afloat in the middle of the fjord, with moorings and a platform and cylindrical buoyancy devices underneath but disconnected from most other things that make up a human community and accessible only by boat. We did all this so that the salmon can remain where *they* need to be, in their water world, contained within the netting, and hence accessible to us even if they are mostly out of sight.

The entire salmon domus could be seen, then, as a complex material interface, a set of textures and *affordances* that serves to negotiate insides and outsides, defining, situating, and/or mediating across various boundaries.[9] One such boundary is the water surface, signifying the two mediums that laws of gravity hold separate along a horizontal plane. It doesn't need any help; it pre-exists practically everything on earth, including the species whose evolutionary development it shaped. The domus then, is situated right "at the edge," on the water surface, from where it allows further practices of separation, mediation, and transformation to occur.

Another surface is the submerged netting that softly defines the edge of each pen or cage, outlining the limits of salmon movement for the time being—and often of wrasse movement too. As we have seen, it is porous and highly selective, allowing free flow of water and marine parasites like sea lice and an occasional young pollock as well, while keeping larger fish and schools of mackerel, cod, and

FIGURE 3. Cages with salmon jumping (gantry up above) (photo by John Law)

pollock out.[10] But the netting fibers allow other affordances as well. Together they offer a structure on which algae may grow, attracting and building relations of a very different kind. The gradual growth of algae, referred to as biofouling, slowly reduces the permeability of the netting. Hence, net permeability is not a given but gradually diminishes as the algae attach and begin to grow. One way to deal with this is to leave the netting to dry. At Vidarøy, this is done by allowing each cage about twice the length of netting that is required. At any given moment, half the netting is rolled up on a cylindrical device out of the water, to dry. The excess netting can be rolled in and out of the water as needed. This process, referred to in Norwegian as *tromling,* is part of regular maintenance work.

In this way, the netting, a seemingly passive tool of separation, is itself enrolled in a wider meshwork of opportunistic life, or, as Anna Tsing puts it, it is "part of that design, intentional or unintentional, that gesture[s] towards the future, making worlds for the yet-to-come, as well as for the present" (Tsing 2013, 28–29; Haraway 1988). Although Tsing refers to critical descriptions of living things, I cite her here to push at the boundary that is thus established between living organisms and those that are passive, like rocks or netting fiber. I want to emphasize that while the distinction can be a valuable heuristic device, it does not always work in practice.

The algae attached to the netting remind us that the domus is a heterogeneous assemblage consisting of matter and materials both dead and alive, both human and nonhuman, and that the distinctions are not always clear-cut, or even relevant. The work that the netting performs unfolds instead along the temporal ebb and flow of biofueling (algae growth) and human labor (*tromling*). Each of these heterogeneous practices relies on and enrolls either netting fibers or the sun and wind in order to achieve their opposite futures: a netting that is either lively with algae, relatively impermeable and incompatible with salmon growth, or relatively "dead," permeable and conducive to salmon growth and well-being. This is how the netting becomes part of its wider assemblage and is, as the domus itself, in a continuous state of becoming.

Drawing on literature on terrestrial domestication, we may identify the netting as a tool of spatial confinement, comparable to fences in a field or stalls in a barn. The analogy is not entirely wrong, but it is incomplete, as it fails to see all the things that the netting *also* is, while selectively ignoring the many ways in which the notion of confinement fails. It is true that it mostly holds salmon in place (though escape accidents do happen; see chapter 7), but to see this as an essential and defining feature of domestication is to partly miss the point. As this and subsequent chapters will show, simply holding salmon confined within a pen is no guarantee of anything and certainly is not enough to make salmon thrive and grow.

The heterogeneous assemblage that I have referred to here as the salmon domus is intimately connected with what farmed salmon are and what they may become. The salmon domus invites and allows particular human modes of being, of movement, of sensing, and of knowing salmon. If domestication is a process of mutual becoming, then it cannot be conceived of without acknowledging the many different affordances and interfaces that constitute the domus as an assemblage of materials. In this section, I have introduced a few of these: the water surface, the netting, and the fluid medium of water. In the next section, I will focus more specifically on the human-salmon relation, exploring practices of (mutual) knowing and responding.[11]

PARTIAL AFFINITY AND EMBODIED COMMUNICATION: RELATIONALITY ACROSS THE WATER SURFACE

Domestication often involves some kind of proximity. It can involve confinement and control, but it can also involve mutuality and offer possibilities for embodied communication (Despret 2013).

In our house, we keep a dog. She is known in my family by the name Laika. Like most golden retrievers, she is obsessed with food and thus easily trained. There is no doubt that she knows me. She also knows that I often wear slippers at home. Together we have invented a ritual that we repeat every time I return home. When she greets me at the front door, I ask, "Where are my slippers?" and she turns around and searches the house. After a while she returns to the kitchen with one slipper, and then I ask her to fetch the other. She races off with a growl of excitement, returns again, and drops the second slipper on the floor. I reward her immediately with a tiny piece of stale bread. She is delighted and expresses great enthusiasm every time, and so do I (even if I find it a bit tiring sometimes), thinking that I cannot ever let her down. Perhaps she feels the same way. Or perhaps she is just desperately hungry.

Salmon are hungry too. With feed pellets I can attract their attention. Below is an excerpt from fieldnotes I made a year later. The site was the same, the crew was more or less the same, but the fish were different: all of last year's salmon had been slaughtered; the site had been fallow for six months and had only recently been reinhabited by a new cohort of smolts, destined for slaughter in 2011. On this visit, in June 2010, they weighed less than 300 grams and had recently been transported from the freshwater tanks at the smolt production site onshore, which I will refer to as Frøystad, where they spent their first year and a half. In the meantime John and I had begun to see ourselves as experts at *daufisk* routines, and we took part in most other activities too, such as sea-lice counts and sea-lice treatment, as well as feeding. This afternoon we were doing what the operation manager, Fredrik, calls 2 percent feeding. This means that we literally took about 2 percent of the allocated weight of pellets and fed it by hand, throwing it randomly around the edges of each cage. Why? Young fish that have just arrived often have a poor appetite. The idea is to reach fish that stay close to the edges and are too weak or inexperienced to take part in the scramble for food in the area where the automatic feeders distribute the pellets.

> We feed slowly. Speed is key. One bucket for each cage first, and then a second round. Moving around one side of the square, then the next side, I start throwing half a scoopful at a time into the cage. But Fredrik takes the scoop and demonstrates: with a lot fewer pellets and a swinging movement with his arm, he achieves much greater dispersal, and thus more fish are likely to get a bite.

Feeding is done mechanically, but it is also done by hand, and particularly at precarious moments like this, when young smolts have just arrived. And it is embodied: a certain grace is helpful, as Fredrik demonstrated. I reported in my fieldnotes that

> I like this. When I go slowly enough, I can see the fish. Barely visible now, but I still see them a few seconds after the pellets have sunk a bit, moving shadows. They are particularly active in cage number 6, but also visible elsewhere. In cage 6 I get the feeling that they follow me around as I gradually throw pellets in crescent-like patterns, "swoosh," as the tiny pieces spread out on the water surface; I design a pattern from one corner of the cage to the next. It feels good to throw yet another scoop onto the swarm down there, knowing that they scramble for food, follow it, search for it.

Reading through my fieldnotes, I notice a shift. I had been bored, and now I seem to be waking up. I recall idle afternoons on the platform in summer, when there was little to do and it was pleasant outside. I think this was one of those days. With the hand-feeding, an affective dimension emerges, and I allow it and explore it further in practice and later in writing. I write about cage 6. They are not only growing well and eating well; they also seem to be particularly alert and responsive, at least to me. More fieldnotes from the same day:

> Again, I am intrigued by cage 6. They are so much more visible than salmon in the other cages and so responsive to my feeding them, both today and yesterday. I notice them alongside the edge of the cage. As soon as I place my hand on the railing, they disappear, like a swarm of mosquitoes when you try to hit them, but after a few seconds they are back. I invite them with a few pellets, but the arrival of the pellets on the water scares them too. Or is it my arm, or my sudden movement—I don't know, but off they go. Until just a few seconds later when they come back up again, resuming their previous pattern of movement near the surface.
>
> I feed slowly, moving from one cage to the next. I try various ways of doing this, but what triggers the most immediate response is not the feeding itself, but my movement. I scare them off; that is our most salient exchange. But the second-most salient response seems to have to do with feed. If they are nearby, my slow feeding attracts some shadowy movements—more or less, depending on how I move.

While salmon may be domesticated, they are rarely tame. If taming is a "relationship between a particular person and a particular animal" (Russel 2002, 286), there is nothing of that sort going on here. And yet, there is a specific kind of interaction, not with the particular fish but with the cage, with cage 6, as it manifests as a particular pattern, a way of responding to my presence. If there is mutuality, it is not one-to-one, with single individuals, but with the fish as a collective, like a swarm.

> In cage 6, their attraction is overwhelming, the shadowy movements become silvery bodies, crowding on top of each other, as if scrambling for something, and continuing to do so while I watch, but only as long as I stand almost perfectly still. They are scared of the light and shadow images that constitute the me that they see, but maybe "scared" is the wrong word. Maybe their response is a mere reflex. But they do like the pellets, and that is an extension of me too, in this particular situation. I have never experienced such a sequence of interaction. The way I relate to them by watching and responding to their movement in a simultaneous and spontaneous manner is new. I am tempted to write that it is a bit like a dance, but I am thinking that by "dance" I am stretching it too far.

There are many different registers of affective relationality on a salmon farm. Henry Buller (2013), in an article that comments more generally on industrial farming, mentions two. One is the relation between individual humans and individual animals, which is where we find "the potential for a lively and flourishing intersubjectivity born of shared lives and embodied experiences within the varied material affordances of the farm environment" (170). The second affective register operates at a different scale, that of the mass, the herd, or the multitude. While this may allow individual wonderment, it also conceals: "in their massivity, these herds and flocks become metaphoric and, as such, killable" (170). Buller reflects on his visit to a Scottish salmon farm and states that "this 'mass' of fish—is for us observers an essentially undifferentiated multitude. Bred in tanks containing over a million fry, graded into cages of similar size for smolting and mature growth, the fish are fed, grown and treated as one, a plural trope of productive aquaculture" (156). He goes on to ask at what point these individual fish "break out of their collective noun to become, at last, singular and the objects—or subjects—of our affective relationality" and immediately concludes that it is at the moment of slaughter, as well as in brief moments of handling or intervention that the individual is accounted for (156). Like many other scholars writing about animal sentience and affect, Buller expresses here a concern about the multitudes, the flip side of which we might describe as a fetishizing of the singular, individual animal as a sentient subject and an object of human affect.

A brief visit to a salmon farm can be overwhelming. The sheer numbers make your head spin, and the way the salmon are nearly always out of sight makes it easy to assume that the "city of fish" is a cold, emotionless, industrial machine with no room for affective care, and certainly not for "flourishing intersubjectivity" across the species barrier. As I have tried to indicate, this is not necessarily so. But to know it differently requires ethnographic presence: presence not only in the abstract, under the guise of the presence of the scientific observer, or what Despret (2013) calls a "disembodied body,"[12] but as bodies that respond and explore—that is, "feeling/seeing/thinking bodies that undo and redo each other, reciprocally though not symmetrically, as partial perspectives that attune themselves to each other" (51). Such presence takes time, and is facilitated in our case through engaged participation in daily routines, such as feeding.[13] It is through my repetitive movement from one cage to next, day after day, that I get a sense of the cages as being different. And then I find that there is no need, as Buller suggests, for the fish to break out of the collective to become the objects—or subjects—of our affective relationality. On the contrary:

> I am thinking for a moment that perhaps this is as close as we'll ever get. But then I realize that it is not true at all, because I have held them, felt them, killed them, smelled them, and been physically close and interacting in a much closer way many times before. So what is the difference, what triggers this feeling of interacting? Is it that they are many? They certainly are more than one—my previous close encoun-

ters are with individuals: many, yes, as in vaccination, but always in a sequence, one after the other. The animals I deal with today are not many; they are more like a swarm, a new amorphous entity that is patterned in a particular way and "speaks to me" collectively, by doing a particular pattern: of crowding around the pellets, or swiftly disappearing.... The swarm evokes an experience of interacting that they, as sequenced individuals, do not trigger.

In addition to Buller's distinction between the relation to the singular individual and the mass, we could add another one: between the multitude moving as a school, or a swarm, of its own accord in three-dimensional space and the multitude as many, as numerical sequence. During a day of vaccination, thousands of salmon passed through our hands. While we held each single one, it was hardly a moment of individual fish breaking out of the collective:

> On the vaccination table, they may be wiggling and slippery, but they are "lost," or rather, they have "lost it." They are caught. It is like having a conversation with someone in a coma. Yes, there may be a bodily response, but the Other is incapacitated and barely able to decide its own next move. The fish today, quite literally, were. And so, when they decided *not* to shy away immediately, or when they made themselves visible as a swarm after a moment of fearful flight, they have "chosen," *they have come back*. They can lock me out, but they did not. Not knowing that I am me, they still allow me a glimpse of their swarm for a minute, a sharing of a moment. For me, it is like a reward, and I become grateful. And cheerful. The feeling of excitement vaguely echoes the memory of walking up a hill and then suddenly standing face to face with a reindeer, the moment of stillness before it quietly runs away. I want to show and tell John. I think this is what I mean when I say I have never been so close before.

Rereading these fieldnotes a few years later, I am struck by the extent to which my cheerfulness seemed to depend upon attributing agency to the fish. If so, perhaps this is similar to what triggers the tendency for individuation of animals in animal studies. Surely, an individual animal can look you in the eye in a way that a herd or a swarm cannot. But as these excerpts have shown, there are other ways for multitudes to participate in embodied communication. Empathy seems to be too strong a word for these encounters, too burdened with notions of shared intersubjectivity. Instead, I suggest that we think of them as instances of what Despret calls "partial affinity," a creative mode of attunement that is embodied but never complete.

To say that salmon farms are breeding grounds for affective human-fish relationality would be an exaggeration. The point I wish to make here, as part of an introduction to the salmon domus, is simply that the possibility is there, expressed here as what I prefer to call partial affinity. Most workers have little time for, or limited interest in, the kind of experiments I engaged in on this lazy afternoon. But then, most workers spent a lot more time with salmon than I did. In the following chapters, and especially chapter 6, I will discuss other ways in which affective registers are mobilized as people work alongside salmon in practices within the heterogeneous affordances of a salmon farming assemblage.

But first, let us take the cues from my fellow workers: how do *they* engage with salmon?

THE WATER SURFACE AS INTERFACE: BECOMING HUNGRY

Fredrik watched the salmon too, but he did it differently. After a while, John and I noticed a particular way of seeing, a way of watching, that all the workers engaged in on a regular basis (see also Law and Lien 2013). The practice even has a name: *sjekke foringa* (to check on the feeding) is the vernacular term, and it was done typically three or four times a day. We are back again in 2009, and the salmon are ready for slaughter:

> I see Karl up above; he has climbed up to the narrow gantry above the platform, where he gets a better view of the cages. He carries a bucket and throws pellets onto the surface with a scoop, then he watches their response. He tells me afterwards that at cage number 6, the response was very good. They still have a good appetite, we can continue to feed.

The following day I trail Aanund as he goes out for the bucket to check on the feeding. What does he look for? How does he do it?

> Aanund picks up a bucket and fills it with pellets from one of the bags in the storage room. The bucket is heavy. He picks up a large plastic scoop as well. We begin at cage number 1. The automatic feeding system is running in the background, distributing pellets on the surface in a small circle. Aanund throws a few spoonfuls of pellets at the water surface, at a different spot, while we both watch intently. On the surface there is a certain activity, a constant simmering, like hot water almost at a boil. Occasionally, the surface is broken by a swift movement of silver in the air, visible only for an instant. If I concentrate, I get a glimpse of the firm body of salmon, or its tail as it breaks the surface going down. Beneath the sunlight and the sky reflecting on the surface, we get a sense of many more silvery shadows moving just under the surface.
>
> Aanund doesn't talk a lot. We move on to cage 2, and then 3, 4, and so on, and his comments are sparse. "It is good today," he says when we are done at cage 3. And then, when we move on to a different cage, he may say, "This is not so good." But what does he look for?
>
> When I ask, he explains that it is about how fast they come to feed. While there are nearly always some hungry salmon near the surface, he looks at how many they are and how hungry they seem to be.

> In the end he concludes that cages 3 and 7 were really good. He adds that cage 7 is nearly always good. Cage 8 is usually better than cage 3 (but not today), but not as good as 7. But pen 11, near the end, is no good at all. Fredrik is right. Something is wrong there.

After a few days, I learn to notice some differences. But I am never sure about it, and I never step up to the role of taking sole responsibility for this particular task. Far from being a passive observer, the person throwing out pellets on the water surface is actively engaging in one of the key practices of salmon farming, a practice that requires on-the-job experience and more than just a couple weeks' training. Seeing involves "skilled vision," a multisensorial and tacit way of knowing, one in which learning is inseparable from doing, and both are embedded in a context of practical engagement with the world.[14] Following Aanund, I recall similar moments in Tasmania, where feeding was done only two and three times a day and where a similarly engaged watching was involved in order to decide exactly how much to feed or when to stop (see Lien 2007). In Norway, feeding is done on and off almost the entire day, and there is more concern about feeding too little than about feeding too much. Nevertheless, the engaged observation of feeding behavior is no less important and is used here as one of the main indicators of how the salmon in each cage or pen are doing. The point, then, is not so much to see whether or not they feed but to spot a difference: Are they doing differently? Are they doing better or worse than yesterday? Are the cages different from one another? And if so, why would that be?

Many things can interfere. If, for example, the sun is high, or the water temperature is warmer than normal, then the salmon tend to be farther down in the pen, and it could appear as if they are not hungry. One solution could be to drop down a camera and observe how they feed at, say 10 meters deep. But this would be a time-consuming operation, so unless they are really worried, the workers at Vidarøy normally don't bother. Instead, they engage their own memories of yesterday and the days before and take the weather into account, along with any other information they have about activities that might have interfered.[15] "Checking on the feeding" produces locally and temporally situated knowledge in an environment littered with information. Hence, it takes a trained eye and quite a bit of experience on site to transform the visual image of a simmering water surface into an indication of "a difference that makes a difference."

I never saw anyone writing this down. It was just part of what the workers did during the day. What they had seen on their daily rounds informed discussions during lunch, it was mentioned in discussions with the vet, and it became just one among the many day-to-day practices that constitute what is, at any given moment, "known" about salmon. As a daily practice, "checking on the feeding" becomes a bolt in the dynamic, ongoing "construction site" out here in the fjord, a coconstitutive practice in the enactment of what specific salmon *are* at any given moment as well as a reference point for future decisions. In this way, it also informs what salmon become.

More fieldnotes, from about a week later:

> It is late morning, and Kristoffer has just arrived. Kristoffer is head operation manager for a group of four production localities, including Vidarøy, and he comes out to see us every day this week to make sure things are going well. Much is at stake this week, with salmon ready for slaughter and Fredrik now on summer vacation.[16] Kristoffer immediately starts checking on the feeding, and I join him. After a while John joins us too. We walk down the platform aisle. He throws pellets on the water into cage 4, and he points out how it is "boiling" on the surface (*du ser det kokje*). When this happens *while* the feeding is on, then it is a clear sign of good appetite, and you don't need to check again for a while. Appetite varies, and if it is not good in the morning, then that is a bad sign, because this is when they are supposed to be most hungry. We check the appetite in 5, 7—all good—and then turn to cages 9 and 10. In 10 there is noticeably less movement on the water surface when he throws out the pellets.
>
> Why is this? Kristoffer thinks out loud: Could it be that they have now grown rather big and so there is less space and thus less oxygen, and now with the rise in temperature that becomes even more acute?[17] If this is the case, they should feed less. It is very important not to overfeed, he says, especially at this stage of the life cycle. (Kristoffer turns to John, and explains, in English now, that it is very uneconomical to feed too much, as feed is a considerable part of the total cost of salmon farming.) Another reason for their lack of appetite could be that they were fed too much yesterday. He says you can never trust the numbers that come out of the computer regarding feeding. It could be that the separating device on the bottom of the silo is faulty, so that it distributes either less or more feed than the figure that comes up on the computer. For this reason, keeping an eye on the feeding is crucial.
>
> We move to cage 9, where the feeding response is somewhat more noticeable than in cage 10, and I ask him about the difficulty of judging this, and he says that yes, it takes experience, but there are also very many ways of doing this. The main point, he maintains, is that this is important, "probably the most important part of the job."
>
> At pen 11 there is very little response when pellets are thrown on the water. I ask whether he thinks this is due to PD, but he thinks not. "They have had PD since last summer, and there aren't that many fish that you can see that sulk, so it could be natural variations, but most likely the combination of high density and high temperatures has an effect today."

As I watch Kristoffer check on the feeding, I think that none of this is new, or unique to animal husbandry. If animals matter to people at all, and if it is our responsibility to contribute to their health and well-being, then the questions that these salmon farmers ask themselves are more or less the same ones that people must have always asked themselves regarding domesticated animals: "How are you doing today? Are you healthy? Am I treating you right?" And then, when something is wrong: "Is there anything I can do differently to make it better?"

The questions may be old, but the way they are asked of salmon is new. I can touch my dog; I can feel her ribs or feel the fat between her ribs and my fingertips and decide whether to put her on a diet. I can lift up my horse's hoof if he limps and check to see if there is a wound somewhere. I can walk through a herd of sheep and look for the ones that are sick or not doing so well. With salmon it is different. Beyond touch and with numbers in the tens and even hundreds of thousands, none of these practices will work. What we have instead is a set of practices that render them visible cage by cage and known to us by their feeding behavior. Through such practices, farmed salmon are, quite literally, *enacted as hungry*. It is with the pellets that I can invite them to come towards us, towards the surface. It is as if their appetite replaces our sensory touch, our careful examination of the individual mediated now by a scoop of pellets. And it is the appetite of the random few, those that happen to feed near the surface, that we engage with. We see a snapshot of salmon activity, but this is only a partial glimpse of a whole that is by definition always out of sight, fluid, and never fully known.

In chapter 4, we will learn more about feed, appetite, and growth as statistical configurations. For now, the point is simply to note that pellets are an indispensable link mediating between humans and salmon. If the water surface marks the boundary between our preferred mediums, water and air, then "checking on the feeding" could be described as a practice that transcends, or overcomes, this boundary. But I think a better way of describing it is to say that through the practice of checking on the feeding, the water surface is literally enrolled into the human-salmon relation, transformed into an interface capable of yielding specific kinds of information. The water surface becomes then no longer something that separates, but a surface that mediates, or connects, and an active constituent of the salmon domus.

THE SALMON DOMUS AS LABORATORY: BECOMING SCIENTIFIC DATA

Knowing salmon is a collective practice. It involves people as well as mediators and consists of a broad range of practices, including the ones we have seen so far: throwing pellets, watching, jotting down numbers, inscribing, submitting, reflecting, telling stories, calculating, calibrating a computer, and writing a text. Some versions of salmon are compiled so that they can travel. Other versions stay more or less put and circulate among the crew. Now and then there are visitors, called

FIGURE 4. View from gantry (photo by John Law)

upon because their contribution to the collective practice of knowing salmon is seen as particularly important or because their knowledge, practices, or networks can make a difference in ways that the farmworkers' practices cannot. The local veterinarians are an example of the latter. With district offices located near the major production sites and with specialized expertise in aquaculture, the local vets are often called upon as consultants in times of trouble, such as in the case described next. At other times, they arrive unannounced, for regular checkups.[18]

Fredrik had hoped she would arrive in the afternoon, but Mari arrives the following morning. She is in her late twenties, about the same age as Fredrik, and they seem to get on well. She has been here before, as she visits many aquaculture production sites in the region from time to time. Her first hour at Vidarøy is spent inside, looking at numbers. Fredrik has made a printout of this week's feeding report and the *daufisk* report, and she studies both carefully. Then she asks for more: she wants the sea-lice counts for the last month and a few other monthly reports. While she looks at the papers, Fredrik tells her about pen 11 and what he sees as the PD problem, as well as the high counts of sea lice in that pen. They talk about feed, and Mari asks how much they feed now, specified as the "outfeed percentage" (tons of feed fed / tons of salmon biomass). Fredrik replies that last week it was 0.65, but this week it is lower, more like 0.5. She asks if they have measured average weight lately. Fredrik replies that it is 4.6 kg in pen 11. What about *daufisk*? Fredrik replies that they've had 50 *daufisk* a day, on average, in pen 11, but much less in the other cages. They discuss a bit and agree that for a pen that holds as many as 180,000 fish, the number of *daufisk* is not alarmingly high.

So what are the symptoms in pen 11 that Fredrik worries about? He explains that he sees them floating quietly along the side of the pen, not as mobile as they should be, and some seem to eat less. The expression he uses is *de sturer* ("they sulk, or mope").[19]

"How exactly are they positioned—with or against the current?" she asks. Fredrik is not quite sure, but he thinks they are mostly positioned against the current. Mari replies that this could indicate *PD-sturing*, which is different from *IPN-sturing*.[20] With PD, they float quietly against the current, horizontally and fairly near the surface of the pen, while with IPN, they are almost vertical in the water and wiggle. It is easy to tell the difference, she explains, once you have seen it. Fredrik adds that there is also foam on the water surface, which *might* be feed residue—perhaps he is feeding too much. But he also speculates that it might be caused by PD-affected salmon gulping their feed, a sign of indigestion. Different interpretations are considered, but without a conclusion.

After a while they move outside to have a look at the fish, and for the next hour or so, the conversation continues, in between an occasional scoop of pellets here and there to "check on the feeding." They have a look at each cage, and Mari comments that the fish in cage 7 have a very good appetite. In pen 11, they stop and watch for a long time. There are many questions and not so many answers. What is clear is that some fish are much thinner than the rest, and there *are* clear signs of *PD-sturing*, but the signs are not very severe.

The next task is for the vet to perform an autopsy. There is a small makeshift laboratory in a separate room next to the office. She needs several samples, preferably from pen 11. Outside, we start the *daufisk* routine, but only one dead salmon appears from pen 11. Is there a blockage somewhere? Fredrik comes around, and then Karl, and it seems as though the air pressure is not working right. Very little water comes out, and no fish. Aanund turns on the pressure in a couple of other cages to collect a few more *daufisk* for Mari to dissect, and then I join her in the lab.

> Mari wears gloves and uses a sharp knife to cut the fish from pen 11 open, intestines exposed. Then she lifts up what she explains to me is the liver, cuts it gently, and moves the intestines aside so that the artery along the spine is exposed. Then she cuts the artery with the sharp end of the knife, and says out loud that there is "fibrin" around the heart, which might have caused death in this particular case. Another fish is placed on the dissection table. Mari opens it up. I ask what she sees,

and she shows me the lighter color of the liver and signs of bleeding in the abdominal cavity, both signs of disease. Several dead fish are then dissected, mostly from other cages.

"So you have proof now that there is PD?" I ask.

"No," she says. "PD has been diagnosed here already, on a different occasion. What I saw today are signs of disease that can be compatible with *[forenlige med]* PD, but there is no proof. To get proof, we would need to send samples to the lab. But since PD has been diagnosed here before, it is not necessary."

The discussion between Fredrik and Mari continues as she prepares to leave, and Mari reflects that the PD vaccinations (which they started a few years earlier) are not as effective as one would have hoped, but they may be the reason why so few fish in pen 11 have become sick: the vaccine makes the virus spread more slowly. But then she adds that the PD vaccination can also weaken the effect of other vaccines.

While Aanund takes her back to the shore, Karl and I prepare lunch. In the afternoon, I remove *daufisk* in the remaining cages (5–10); Karl cleans a couple of floating buoys made of cork; and Fredrik drives a forklift inside the feed storage room, emptying the bags of feed into feed line 4, which he has calibrated this morning. Karl points to a school of mackerel, circling the cages on the outside. Perhaps they are feeding off dust from the pellets? Who knows?

Later, after I join Fredrik in front of the computers, Karl calls on his cell phone from pen 11:

> The blockage is now gone—*daufisk* are suddenly gushing out in great numbers; he needs help! With an extra wheelbarrow I rush out and soon take over the work of counting them and wheeling them back to the platform. The blue container fills up quickly, and we have to turn off the air pressure now and then to keep up. Several wheelbarrows later, and with increasing soreness in my triceps, I finally put down the number on the waterproof-paper form: 92 in pen 11. This is a lot more than the average 50 we talked about this morning. It doesn't look good.

Removing *daufisk* is a matter of hygiene, a matter of care. Counting them, jotting down numbers, and submitting them electronically also helps maintain an updated salmon stock inventory (more on this in chapter 4 and 6). In this way, our daily *daufisk* rounds make us not only undertakers but stock keepers too. But we are more than that: the numbers are also an indication of the state of affairs. Like the practice of checking on the feeding, keeping an eye on the number of dead fish coconstitutes what salmon *are* at any given moment. It is a practice that enacts salmon in particular cages as doing well or not so well. As the examples concerning pen 11 show, the evaluation is often arbitrary, and there is no absolute numerical threshold. But together with other observations, the *daufisk* numbers can be mobilized in support of provisional hypotheses about the state of affairs in particular sites. Hence, our role as undertakers enables our role as caretakers, and the two go together in daily practices, undifferentiated, as part of our routine.

As John and I gradually expanded our fieldwork practice on the platform, we often found ourselves enrolled in tasks that followed the protocols of scientific sampling. I came to think of the salmon farm as more than a food-production site. As we sampled, measured, and reported, our concern shifted from the well-being of a particular cage or pen to a more general concern about getting the numbers right. We found ourselves performing a kind of makeshift laboratory practice, enacting salmon as scientific data in experiments that often extended far beyond the confines of Vidarøy and sometimes even beyond the company itself. Counting sea lice is but one example. Consider these fieldnotes from 2009:

> It is an early morning in July, and Karl and I are off to do the biweekly lice count. I fetch a dip net, the form from Hardanger Fiskehelse Nettverk (this too is printed on waterproof paper),[21] a wheelbarrow with a bucket on top (but I make sure it is not the one we use for *daufisk*), a bottle of Benzoak, two pairs of gloves, a rope, and a bucket of pellets and a scoop. Fredrik has told us to do cage 9 today. Time to start. Karl ties the bucket to the rope and drops it down into the cage until it fills with water. Then he pulls it back up and places it on the wheelbarrow. He adds a lid of Benzoak, which works as an anesthetic, and reminds me that it works on humans too so we both put on our gloves. He sprinkles some pellets by the side of the cage, then he picks up the net, bends over, and reaches for the shadows that assemble very near us now, just beneath the surface. Within seconds, he catches three, and with a firm grip he lifts them up and drops them into the black bucket of Benzoak bath, where they wiggle and splash for a little while. "They are hungry today," he says, indicating that that is when they are easy to catch: not-so-hungry fish are almost impossible to catch. After a while the fish calm down, and we take turns picking them up by the tail, holding them head down in the water (they are calmer that way), and start counting. It was difficult to spot the lice in the beginning, but I am starting to recognize the *hoa* (mobile females) as the ones with the long white tails, the *bevegelige* (mobile males) as round, somewhat smaller, and with no such lines extending from their bodies. We see no firm ones (sexually immature lice [*faste*]). We both say the numbers out loud—"1 *hoa*, 2 *bevegelige*"—and then the one who isn't holding the fish writes the numbers on the form. We repeat this procedure until we have done twenty fish, which is the required sample. According to instructions printed on the form, we should sample from the worst cage and pick out the largest fish there. In the end we try to count how many lice are left in the bucket. Karl pours the water slowly into the wheelbarrow, so that we can see the lice, and he notices a few in the bucket too. Even though I do spot one or two, I find it almost impossible to see anything, and think the final number that we added in the end must be rather uncertain.

Counting lice gives us a sense of how our salmon are doing, but that is not the only reason, or even the main reason, that we do it. The information is going elsewhere.

We are participants in a regional system to counteract what was, at the time, a growing concern regarding increased prevalence of sea lice, so the results of our biweekly sample will be assembled with similar results from elsewhere to make up a numerical assessment for the region.

Sea lice move with the currents from one locality to the next. On their way, they encounter smolts on their way to the ocean or returning salmon headed upriver to spawn. This is now considered one of the most important ways in which salmon farming can harm the wild salmon stocks; hence, a concerted effort among the local salmon farmers is called for.

In 2009, sea lice were counted every second week, but the following year, a new regime was implemented in certain regions, including Hardanger. Sea-lice counts were now done weekly, and the results submitted directly to the local authorities in Hardanger, who—in collaboration with the industry and through the Hardanger Fiskehelse Nettverk—had launched a project called Lusalaus ("liceless"). In addition to more frequent sea-lice counts, these measures include a system of zoning,[22] as well as a stricter regime for mandatory sea-lice treatments, with most thresholds indicating a need for treatment lower than the year before.[23]

Sea lice are difficult to spot, and 5-kilogram salmon are difficult to hold. Our precision as contributors to the statistical survey may have been below par, but we did count what we saw. Hence, as we submitted our numbers, we contributed to enacting not only Vidarøy as a salmon domus but the entire fjord as a watershed with sea lice. Together with many other farmworkers elsewhere, we contributed in documenting the need for a stricter regime the following year. We contributed in performing the Hardanger region as a precarious watershed, and farmed salmon as the vectors in need of control. While "checking the feeding" is a coconstitutive practice in the enactment of the state of affairs for farmed salmon in every single cage at Vidarøy, the practice of "counting sea lice" enrolls farmed salmon as statistical samples only and uses these to enact the state of affairs in the entire fjord with regard to the prevalence of sea lice. This too serves as a reference point for future decisions, but these decisions are made not by the company but by the local authorities, assembled in this case under the name Hardanger Fiskehelse Nettverk.

The image of Vidarøy as a house and group of cages gathered on a platform in the middle of the fjord invites an image of the grow-out site as a salmon domus in and of itself. The sea-lice example reminds us that the boundaries of the domus are highly porous. An entire watershed is folded into each single number of *bevegelige hoa*, or "mobile shes" (a term that sounds as awkward in Norwegian as it does in English). Just as the sea lice move with the current and know no boundaries until they reach the river mouth and salinity drops, the salmon domus has no defined inside and outside, but shifting attachments, presences, and absences that make a difference beyond the confines of the locality.

There is a certain comfort in numbers. Even if Karl and I were slightly inaccurate, it is tempting to imagine our submission of these numbers as a contribution to some kind of control—that the most important challenges in the region are properly accounted for and effectively handled. To some extent this is the case, and the policy is sincere enough. But it is important to differentiate between a sincere intention of control and control as a resulting outcome. In January 2010 I had the chance to discuss the newly implemented sea-lice measures with a senior vet in the region. She explained the new regime and how the zoning would require quite a lot of resources and practical adaptations for the salmon farmers involved. I asked if she thought it would help. She replied:

> It is ambiguous. It could also mean greater numbers [of sea lice] in particular areas at particular times. Another measure that might have worked even better would be to keep the fish at sea for slightly shorter periods [i.e., keep them for longer periods in tanks onshore] so that only, say, half the fish were in the sea when the smolts came by. Another important measure which is implemented already is to avoid moving fish at sea, and to separate them according to size before slaughter. . . .
>
> Some of these regulations are characterized by a lack of knowledge. The thresholds for treatment, for example, should be more differentiated. For instance, in summer, when smolts are no longer around, they could be higher, because then the wrasse are fairly successful. But in order for wrasse to do their job, they need something to work on! When the thresholds are tightened and medication enforced instead, this is no good because pharmacological treatment breeds resistance—it is pure mathematics.

Four years later, the numbers indicate that her worry was not unfounded. In 2014, the Norwegian Food Safety Authority reported that the medication against sea lice has become less effective and resistance seems to be on the rise. The use of sea-lice medication in Norway increased from 5.516 kg in 2009 to 8.403 kg in 2013.

"A FRAGILE MIRACLE"

Only forty years ago, there was no industrialized salmon farming—in Norway or anywhere else. While other husbandry animals have evolved with humans for millennia, salmon are indeed "newcomers to the farm." Not surprisingly, we wonder about many things. We wonder not only about the salmon. We wonder about the feed pipes, the calibration, the temperatures, the netting, the generator, the sea lice, the air pressure, the blockage, infectious diseases, and much more. The entire assemblage is what John Law calls "a fragile miracle," a place where anything can go wrong, a place where there are so many *different ways* in which things can go wrong, and where the work of holding it together is all about practical, mundane tasks like checking, repairing, pondering, worrying, supplying, counting, and reporting, or as in the cases above, implementing a zoning regime or calling the vet.

Just as the domestication of husbandry animals irreversibly altered the bacterial and viral environment of humans, and hence our human immune system, the domestication of salmon has microbial side effects too. Sea lice thrive, as we have seen, but new or previously unknown viral and bacterial diseases have appeared as well and have been identified, classified, and domesticated as objects of research and sometimes have been nearly eradicated due to effective vaccines. Assembling thousands of salmon in a confined locality causes changes in the microbial environment that are hard to foresee.

During the early years of salmon farming, bacterial diseases were rampant, and antibiotics were the only remedy, so they were used in large quantities. Today, the spread and prevention of contagious diseases is better understood, vaccines have been developed, and the use of antibiotics, especially in Norway but also in Canada and Scotland, has dropped significantly as a result.[24]

Disease outbreaks and parasite attacks are just two ways in which the salmon assemblage is fragile. Holding it together requires constant practices of knowing, caring, and tinkering, and mobilizing networks that extend far beyond the salmon farm. Salmon farming is a precarious practice in which humans are active participants but are hardly in control.

In this chapter, I have introduced a few of the practices that call for human attention at Vidarøy. Focusing on these practices, I have tried to show how humans never act alone—that the practices are *heterogeneous*. Nets, buckets, knife, pellets, waterproof paper, Benzoak, and the salmon themselves are all active, though not always fully trusted, agents in holding this fragile assemblage together. And yet, this is just a fraction of all the things that an operation manager needs to attend to. I will conclude this chapter by taking a long last look at the to-do list, for week 27, posted on the wall in the Vidarøy office (my translation):

- Send weekly report to Henrik, Hans, Helge, Kristoffer
- Call Jon about color samples (all pens?)
- Do an extra sea-lice count
- Send remote control to Maritime Elektro A/S
- Count extra hours (vacation)
- Fetch more diesel at the harbor
- Check new diesel tank
- Calibrate feed silo 1 and 2 when they are empty
- "*Tromle*" all cages with even numbers
- Order feed
- Count and make an overview regarding the generator filter
- Check wrasse covers
- Wash and disinfect lightbulbs
- Place cage 6 "on hunger" (*på svelt*), await further notice from Helge
- Take feed samples

All these tasks are in addition to carrying out the daily routines, sending daily reports, doing small repairs, and calling for the vet: What goes on the list are things that need to be remembered, and while some may be regular, repeated tasks (*tromling*, feed samples, weekly report, order feed, fetch diesel), they are still written down so that the tasks may be shared and not forgotten. Some items are about accommodating the immediate environment, such as checking and replacing the wrasse covers and cleaning the netting (*tromling*.) Many items involve the maintenance or repair of other things, such as the lightbulbs, the diesel tank, and the remote control. But the most time-consuming items for Fredrik are not even listed here, such as the morning he spent with the vet and the hours he spent on the computer, sorting and submitting various kinds of information electronically, according to specified formats and recipients. Finally, there is uncertainty. There will be slaughter, but we don't know exactly when. We need to do color samples, but where exactly we need to do this depends on what the managers over at the head office have to say.

. . .

It is 3:30 in the afternoon on the last day of June, and the automatic feeding just turned itself off. Gradually, as the four feed pipes shut down, one after the other, the platform becomes quiet, and all we hear is the sound of seagulls up above and the reluctant engine of an outboard motor. Aanund is starting the boat already.

"18,850," Fredrik shouts.

"What?"

"That is how much we fed them today. It is pretty good, I think."

The number, which refers to kilograms, has already been submitted and has arrived, electronically, in the offices onshore where we are now headed. It is a number along with many others with which salmon is made to travel, from the platform in the middle of the fjord to the head office, which in more ways than one "holds the firm together." This is the topic of the next chapter.

4

BECOMING BIOMASS

Appetite, Numbers, and Managerial Control

The salmon domus is a fragile assemblage, and fish are lively beings. Even in confinement, their fleshy vitality resists human attempts at control. But they can still be managed. One way of making the salmon assemblage manageable is through numbers. That is what this chapter is about.

The salmon domus can be conceived as a locality, a site. In chapter 3, we explored some ways in which the salmon assemblage can be conceived of as a site of domestication. I described the salmon domus as it unfolds at Vidarøy, a grow-out site in the middle of the fjord. But the salmon's lives don't begin there, and neither do they end there. In this chapter, as well as in chapter 5, we will explore the domus at different moments during the salmon's life cycle. We will weave back and forth between salmon at different stages of growth and look at how their enrollment in the salmon assemblage inscribes and enacts particular temporalities. We will explore the translations that enact salmon as commodity as well as profit and that connect the livelihoods of people in Hardanger with the appetite for salmon in other parts of the world.

The salmon domus is also a temporal assemblage. The entire operation could be described as multiple, repetitive practices of cleaning and feeding, as well as of the repeated shifts between work and leisure that are regulated by Norwegian labor laws. There is the temporality of the salmon life cycle, which is loosely coordinated with a spatially distributed structure of localities, each of which is designed to support the salmon's needs at a particular stage of life. Hence, hatcheries are different from smolt production sites, which differ again from the grow-out sites in the fjords, such as Vidarøy. These are, to some extent, managed as separate economic units, and indeed the different worksites are social units too, each with its own social atmosphere and permanent staff. Because salmon at different stages of the life cycle (fertilized egg, alevin, fry, parr, smolt, and finally seawater salmon) have different needs, the growth and development of batches of salmon involves, and is also registered, as a move between localities. Each move is accompanied by a written form that details the life history of the specific batch of salmon thus far, to ensure traceability and settle any potential future disagreements or complaints.

And then there is the temporality of the firm itself: its annual budgets, profits, Christmas parties, internal newsletters, weekly meetings, and growth figures. And if we zoom out a bit more, we may note the cycles of Norwegian regulatory authorities, manifesting through their requirements of annual, monthly, or weekly reports of figures regarding taxes, profit, biomass, escapees, and lice. Sometimes these multiple temporalities are closely aligned, but often they are not. Much of what goes on in the industry is about establishing coherence, assembling the various temporal cycles at particular moments, so that as a result the specific locality, the firm, the region, or the Norwegian industry becomes legible as a coherent whole. In chapter 3, such practices were described as inscriptions: as the writing down of numbers on a waterproof sheet of paper or entering figures on a computer offshore. In this chapter, I shall trace the journey of some of these numbers and show how they become significant as agents that enact the firm as a coherent entity. Such practices subsequently lay the foundation for further capitalist expansion, which is the topic of chapter 5.

A GLOBAL BULK COMMODITY

Sjølaks is an enterprise designed to produce a profit. Practically all of its salmon is exported. Every Friday morning, the sales director spends several hours on the phone at the head office, negotiating salmon deliveries for the following week. At the other end are exporters located in offices in Bergen and other towns in Norway, with online connections to importers in Japan, China, South Korea, South America, France, Germany, and other countries around the world. The negotiations are all about price. How much are they willing to pay for a shipment of Sjølaks salmon? How much can the exporter make on the transaction this week? For Sjølaks, profit is the marginal gain calculated as the difference between the cost of production per unit of weight and the sales price per unit of weight. Every extra gram of salmon flesh added to each fish during the production cycle adds to profit, provided that the cost of feeding does not exceed the extra gain per unit weight. Large salmon are more profitable not only because they mean more salmon flesh sold in total (expressed as higher gross revenue) but also because the cost of production over the entire salmon life cycle can then be divided by a larger denominator.[1] At Sjølaks, which is a typical medium-sized company, feed constitutes about 50 percent of the total cost of production, and salaries constituted around 16–17 percent during the time of our fieldwork. Analysis of Norwegian salmon farming indicates an average profit margin gain of around 30 percent in 2006.[2] Since then, the cost of production has gone up, due primarily to the increased prevalence of sea lice, but because global salmon prices have gone up as well, the profit margins for most Norwegian producers have remained high. This was also due to the collapse in the Chilean farming industry in 2009,[3] which reduced the overall supply of salmon on the global market for several years.

An important strategy to increase profit margins in food production during the past decades has been "value adding," or enhancing some aspect of the product so that consumers are willing to pay more for the same unit weight of food purchased. This is particularly important in so-called saturated markets (see Lien 1997). Norwegian salmon producers explore few such opportunities. Instead, they tend to see profit as a relation between cost of production and global prices, the latter being volatile and largely beyond a firm's control. This can be partly explained by the system of distribution and the way Norwegian farmed salmon is transacted globally as a bulk commodity. Unlike other global export products—for example, wine and cheese, which are priced according to a sophisticated scheme of quality differentiation[4]—farmed Atlantic salmon is priced in a straightforward manner, according to the grading criteria defined by the Norwegian Industry Standard for Fish.[5] According to this standard, farmed Atlantic salmon are graded in three categories: Superior, Ordinary, and Production.[6] In addition, there is a certain price differentiation depending on size, reflecting some variations in size preference across different markets. In Eastern Europe, for example, where salmon is often baked, there is a preference for smaller salmon. Where salmon is more often sold as steaks, a preference for larger salmon is more common.

This simple system of differentiation allows for a similarly simple monitoring of salmon prices worldwide. So-called spot prices are publicly recorded on the basis of actual sales and form the baseline for the negotiations between the sales directors, the exporters, and the importers at the other end. The prices fluctuate and represent potential gain as well as risk for the producers, whose investment follows the salmon life cycle and is thus necessarily long-term.[7]

For a producer, then, the aim is to produce as much "Superior" salmon as possible, of sizes that fetch a decent price, which is mostly around 4–5 kg or more. Most batches are a combination of Superior and Ordinary, and the proportion of each reflects aspects of production during the entire life cycle, including the slaughtering process itself. As a general rule, healthy fish tends to qualify as Superior. But since the quality can be determined only after slaughter, quality becomes in practice a way of adjusting the final payment. This implies that there is limited scope for the firm to increase profit through quality enhancement beyond ensuring a viable and healthy batch. Value becomes instead a relation of fluctuating prices in relation to weight. The concern for the firm, then, as truckloads of salmon are about to be shipped from Hardanger, is not so much its nutritional value, its mode of production, or the distinct flavor or color of its flesh (although the latter is of some importance). What matters more is *how much* salmon there is and *when* it is sold in relation to highly fluctuating and largely unpredictable spot prices.

Hence, from an economic perspective, the main difference between the first-feeding fry and the salmon that are shipped off for slaughter about three years later is, literally, about 5 kilograms of salmon flesh. What goes on in the meantime is geared towards enhancing the total amount of healthy salmon biomass available for slaughter. This means not only ensuring steady growth but also averting crises and costly labor operations. This is hard work, it is serious, and it is monitored very closely and revolves around numbers. The point was brought home during a casual conversation I had with an operation manager from a grow-out site further north. Over a glass of beer, he reflected on the Monday morning meetings, when all operation managers gather to defend and discuss their figures from the previous week, which are reported in the weekly report:

> It is fairly tense. Everyone has to defend their numbers, whatever they are, good or bad. The owner is there, and the director of finance too. But I think that none of us [operation managers] take pleasure in other people's problems, because we all know that next time it could be us. I remember my first Monday morning meeting many years ago. I was new to the firm, and Henrik made it clear to everyone that my forms were full of mistakes. Then he took the sheet of paper with my weekly report, folded it into a little ball, and threw it into the wastebasket, just to illustrate how useless it was.

So what is it about numbers? What are they made of and why do they matter? In this chapter we look at the different practices that make the salmon assemblage legible, visible, and manageable as an economic entity. It is important to keep in mind that it is as parts of *an economic unit* that places like Vidarøy are sustained over time. The workers are there not to grow their own food but to secure a monthly salary. Their salaries are negotiated every other year and are paid irrespective of salmon price fluctuations, but their long-term job security depends on Sjølaks's sustainability as an economic enterprise. It also depends, obviously, on each employee's readiness to take on responsibility and do what is needed to make the salmon grow and stay healthy.

A great deal of responsibility for the daily operations is delegated to staff members, who plan their workday relatively independently, without anyone looking over their shoulder. When managers come by, it is often to help out, to consider a problem, or to have a chat. This style of management—characterized by egalitarian social interaction, democratic ideals, and an emphasis on informality—is often held to be a characteristic feature of Nordic business and management (see, for instance, Byrkjeflot 2001; and Brøgger 2010).[8] But this somewhat egalitarian ethos does not preclude the existence of a clearly articulated hierarchy: as the Monday-morning incident with the wastebasket indicates, there is no doubt whatsoever about who is in charge.[9] People *know* where the orders come from, and even if some occasionally voice a different opinion, production policies are implemented through delegation of responsibility. This means that management is not performed by giving detailed orders or by authoritarian enforcement. Instead, it is performed through numbers, or through what we might call an internal audit routine. Let us look at how this plays out in practice.

HOLDING TANKS TOGETHER

> Tore has had a busy weekend. Broadly built and rather quiet, he is often ironic and very funny. But this morning he is tired. Even though he was not officially on duty at the smolt production site—and someone else was on call over the weekend—he still had to get in his car and drive over here to help out. It is not the first time, and certainly not the last, that he spends a Saturday or Sunday between the tanks.[10]

We are at a smolt production site that I call Frøystad. Like Vidarøy, it is one of the sites that we keep coming back to. Frøystad has been in operation for more than thirty years, under different owners, and the site of different experiments in aquaculture.[11] Tore recalls how he did some carpentry work here as a young man in the early 1970s. Now, as the operation manager, he is responsible for more than 800,000 smolts, distributed in sixteen tanks—some outside, others covered by a black plastic roof in what feels like an inverse greenhouse: warm and humid but completely sealed from the midsummer light outside.[12] He is responsible for other people too; Lars and Kristin are permanent staff and nearly always gather with him at lunchtime. But this week there are also two women who work part-time and show up on alternate days to help out with vaccination, which keeps at least three people busy at the same time—and provides an opportunity for John and me to be useful as well (for details, see Law and Lien 2013, 2014). This morning I have joined Tore in his small office, overlooking the blue-green cement tanks outside. I am trying to get a grasp of his paperwork and how numbers travel. But first I need to know what happened here over the weekend.

> Lars was on call, and had been woken up by the alarm in the middle of the night. (The alarm at the smolt production site is connected to a cell phone that is always carried by the person on call.) He got dressed and drove over to the site immediately. The display indicated that the water levels in the tanks had become critically low. But why? It took a while before they figured out that the problem was upriver, at the dam where there is a pipe that supplies the tanks with water. The end of the pipe is fitted with a metal grille, a kind of filter, to prevent debris like branches and leaves from entering the water supply system. Now the filter was clogged up by autumn leaves.
>
> The following day, John showed me what Lars had showed him the day before. There is a wooden ramp, and from under this ramp the pipe, which is about 30 cm

in diameter, runs all the way down to the water-circulation system at the smolt production site. More precisely, there are two pipes: the one below is in use; the one on top is there in case the lower one fails.

Always a backup, John reminds me. He tends to notice these things.

Tore explains: "When there is lots of water in the river, water and leaves flow steadily over the edge of the dam. But when the water levels drop, like now with very little rain, the leaves remain in the dam and clog up the grille, and then not enough water flows through to our water-circulation system." As John and I were learning about hatcheries and smolt production sites, I began to think of grow-out sites like Vidarøy as relatively simple assemblages, places where very little could go seriously wrong. Compared to the complexities involved in securing a steady water supply and the constant need to monitor the water medium in tanks (where a drop in oxygen can kill the fish within a few minutes), sites like Vidarøy appeared fairly simple to run.[13]

Tore tells me that he sometimes gets up at five in the morning to get to the office early in order to catch up on the paperwork. On his desk there is a large ring-binder with sections for the different types of measurements. Each day, numbers are entered on various forms and then carefully placed in this folder. I leaf through it and find records of water temperatures and precipitation going as far back as 1986. And I realize that what they measure is a lot more comprehensive, and has a wider spatial circumference, than routine measurements at Vidarøy. One of the things they collect is water.

> On the shelf in the feed storage room there are water bottles with inscriptions like "7-21" or "6-20." Kristin explains that they collect these daily from the river just below the bridge and leave them on the shelf for a fortnight or so "in case something happens." "Something" could be a catastrophic event with no apparent cause, in which case the collection of water bottles could be part of the forensics. Kristin tells about an incident in February, when suddenly the fish started acting strangely, jumping up on the side of the tank, really crazy and like nothing they had ever seen before. A few months later it was concluded that it was an instance of too much aluminum in the water supply, partly as a result of unusually cold weather, which had caused damage to the silica pump—silica is added to the water to counteract the effect of aluminum.[14]

In the beginning I was surprised at the huge amount of work here that involved monitoring practices and inscription devices of various kinds. The two most important variables are water levels and oxygen. Both of these are linked to the alarm system. The lower level at which an oxygenator (or aerator) kicks in is 85 percent oxygen; and the alarm goes off at 65 percent.[15] But routine monitoring practices are not confined to the tanks or even to the work site, but extend far into the woods as well. More fieldnotes:

Kristin tells me that the other day they followed the river up into the forest, to get to five different sites where they collect water for pH measurements. "A nice walk," says Kristin. "You should go there some time." The farms up there use fertilizer, and with heavy rain the runoff can change the acidity of the water in the river. Pine needles have an effect too. All of this is beyond their control, but it needs to be registered. They do this walk only every month or so.

Every morning they also record the weather, including wind, rain, or snow, and the temperatures of the air and water.

Back in the office, studying the curve on Tore's computer screen that details oxygen saturation in the covered tanks for the previous day, I notice a sudden drop down to nearly 70 percent before it picks up again. Tore explains that this was when they turned off the lights last night. The sudden loss of light made them scramble for the bottom of the tank, and their movement increased their overall oxygen consumption. Such drops in the oxygen curve can be a result of outside sources, but they also happen when the fish get "stressed." But then they calm down very quickly afterwards. He explains that one lightbulb near the tanks is painted black. It is always on to provide a little light; if it weren't there, the reaction would be even stronger.

As we talk, I put together a list of all the things that are measured regularly. The list does not exist in this form anywhere except in my fieldnotes. For Tore and his coworkers, it materializes as various forms waiting to be filled out or as routine practices that are repetitive and hardly need to be written down. Other measurements are done automatically by sensors linked to computers that transform numbers into curves, like the one for oxygen. Here is my list of things that are measured:

- O_2—measured continuously. (An alarm goes off it drops too much; otherwise, when it gets down to 85 percent, it automatically increases;[16] similarly, there is a monitor to bring the percentage down if it goes up too much. This is not done manually, so we don't really see it, but I guess it goes into the general computer.)
- CO_2—measured weekly.
- Water current in tanks—used for the first time today with the arrival of fish, but then probably weekly. The device was new last summer.
- pH at the smolt production site—measured weekly in river near the water pipe and in the water-circulation system and tanks (*vassverket* and *driftsverket*).
- pH further up in river catchment area—measured monthly or as time permits (there are five sites).
- Feed distributed to each automatic feeder—measured daily.
- Morts, or *daufisk* – counted and removed daily.
- Environment (*miljømålinger*)—measured daily (except pH, which is weekly).
    - Water temperature—morning and night
    - Air temperature—morning and night
    - River pH near water pipe—weekly, see above

- Weather—daily[17]
- Wind—daily
- Rain/snow (millimeters)—daily

While the netting holding salmon in place at Vidarøy is permeable, allowing a continuous exchange of current and microorganisms, the tanks at Frøystad appear at first to be sealed off. But the immediate environment—wind, rain, aluminum, and acidity—interact with the water medium, and thus with the fish, in a most intimate way. Contrary to its appearance as closed confinement, the smolt production site reaches out beyond the production site itself, to the river and into the forest, past old farms and farmland to places defined as monitoring sites for pH (as well as to the oxygen supplier). Because it reaches out in so many directions, the salmon in tanks are rendered vulnerable and in need of constant surveillance. I need to reconsider my understanding of a tank. More fieldnotes, from later the same week:

> So what is a tank? It is a round cylindrical structure, which used to be only 2–3 meters across in the old days, but which is now up to 10 meters in diameter, or even more sometimes. The depth varies: the bigger ones are around 3 meters, and the slightly smaller ones outside of the black plastic are probably around 2 meters or a little less.
>
> The tank seems to hold the fish in place, seems to allow some sort of controlled environment, seems to represent some kind of closure (like the entry gate down by the river, with the salmon-made-out-of-wood on the handle and a sign that politely asks visitors not to enter, but to consult the Sjølaks offices across the bridge). But the tank allows between 1,000 and 1,500 liters of water to flush through, fresh from the river, and the impression of closure is only an effect of my own misguided assumption—because the tank is really only a device to hold the water-salmon assemblage upright, stable, and in place, so that it can be dealt with, fed, and monitored. The monitoring is intense, but the tank's ability to keep the surroundings out is less efficient. The outside tanks are covered by nets to keep birds out and to keep the mink out as well (Kristin says they live up the river), but the rain will come through; the sun and the wind and the temperature will affect the tank environment directly, slow down the feeding, decrease or increase the salmon's appetite, or growth. In a way, these tanks are not much more than a carefully crafted pond with exit and entry points closed for the moment, artificial ponds that allow humans to monitor the process of growth towards smoltification so that they can be taken and relocated at the right moment.

To think of the tanks as dynamic entities where environmental features are constantly folded in makes it easier to see the need for numbers and why constant surveillance is crucial. Out at the fjord not much can be done about the weather and the current. One can check water temperatures for the record, but it doesn't make much difference. Here at Frøystad, low temperatures can make water freeze in pipes, the wrong pH can negatively affect the fish, high aluminum can make them go crazy, autumn leaves can clog up the water pipe so that water levels drop, and oxygen can become too high or too low, which can kill the fish. Because it is all inside a tank, any deviation from the norm can be magnified to the extreme, and the risk of catastrophe is immediate and ever present. In a sense it is because the fish are *not* immersed in the sea that Kristin's routine care for salmon leads her upstream and into the woods. If this is "risk society," the risks can be mitigated, and workers are on constant call and responsive to the slightest change in the fish's environment. This makes the smolt production site rather different from grow-out sites like Vidarøy, where the "society" in question would be the entire Hardangerfjord ecosystem, and a worry more for the environmentalists and marine biologists than for Fredrik and Karl. While risk management at Frøystad is largely contained within the organizational structure of the firm, risk management at Vidarøy extends those boundaries. There *is* no simple "alarm" function at Vidarøy, but rather a handful of practical, mitigative, and audit-like measures (such as lice treatment, reporting escapees) that together provide some protection against salmon farming's potential negative effects on surrounding waterways. With no one directly "on call" and no simple solutions available, risk-management practices appear both delegated and less predictable as they are distributed across various layers of regulatory authority.

Yet, in relation to Sjølaks as an economic unit, smolt production sites and grow-out sites are not all that different. The forms that need to be filled out are more or less the same; the numbers that count are similar. Let us return to Frøystad, the smolt production site.

COPRODUCING SMOLTS AND NUMBERS

The list of things to measure at the smolt production site on a regular basis is not the same as the list of things to report. Between Frøystad and the head office, there is a river. The houses are only about 50 meters apart, but we need to walk out on the main road and over a bridge to get there. Although it is a short walk, there is normally no need to go. Everything is reported electronically. The broadband connects head office with all its production sites, near and far, and sometimes the distance seems greater than what it appears to be on the map.

Hi sio is the vernacular dialect expression that workers at the smolt production site often use when referring to the head office. Meaning "the other side,"[18] the term indicates location as well as organizational othering. *Hi sio* is where strategic decisions are made, *hi sio* is supplied with numbers of all kinds, and *hi sio* will now and then give instructions too. There is no doubt that it could be described as a "center of calculation."[19] But *"hi sio"* expresses more than that: it makes it clear that "it" is not "us." Boundary work is going on, enacting insides and outsides. One morning, as John returned to the smolt production site after a brief visit to *hi sio,* someone jokingly welcomed him back from the West Bank. Small talk and storytelling enact this and many other boundaries, such as gender and age. In this way, local gossip, age, gender stereotypes and organizational order merge in daily conversations, sustaining a sense of work ethic and solidarity, as well as a moral economy around which certain norms and expectations can be articulated.[20]

But aside from this, the relation between an operational unit such as Frøystad and the head office is firm and strong. It materializes most predictably as the regular transfer of numbers. And it is sustained by people like Tore, who puts in extra hours before dawn to finish the necessary paperwork. Reporting takes a lot of his time and effort, even though he submits his report only once a month; at Vidarøy reports are submitted weekly. But the parameters are similar, and the procedures for keeping track are more or less the same. So what does a report consist of? Tore lists a few items off the top of his head: "There is the *daufisk* (morts), disease incidents (if there are any), how much feed we have fed, and the FCR, or 'feed conversion ratio' [*fôrfaktor*]. In addition there is the paperwork that accompanies fish delivery, both incoming fish and outgoing fish."[21]

Leaving the latter aside for the moment, let us take a closer look at the FCR. It is calculated as the "quantity of feed divided by biomass gain" and is a measure of the efficiency with which feed is converted to salmon flesh. It can be calculated for a specified time period on the basis of quantities of feed fed, provided that figures for fish biomass are available within a given time period. This means, in practice, that in order to calculate an FCR, one needs to know not only how much feed is fed to a specified unit in a given time period (which is relatively simple) but also the weight gain of the fish.[22] Because the FCR provides a specific measure of the dynamic interrelations between fish and feed, it enacts a specific temporality, one of biological growth. But because it is also an expression of growth *relative to feed,* which is closely connected to cost of production, the FCR is also in practice a measure of cost-effectiveness. In this way, the FCR operates as a mediator between the biological reality of salmon as living, growing entities and the economic reality of Sjølaks as a more or less profitable enterprise. For example, when I asked the finance director in 2011 about the current state of affairs at Sjølaks, he moved swiftly between the realm of production and prices: "The cost of producing salmon—which has fallen steadily over the last couple of decades—has actually gone up now in the past few years, quite dramatically! This is because of the biology *[biologien]*, which has become more difficult. In addition, the cost of feed has gone up, both in absolute terms (feed is more expensive because fish oil and fish meal are getting scarce), and also relative to growth: our FCR has gone up from 1.10 about ten years ago to 1.30 today."

The reference here is to what is called the "economic FCR," which takes into account the losses represented by morts and will thus be affected (increased) with incidence of disease.[23] In spite of this, the company is doing well and investing heavily in new production units. The reason is that the prices of salmon happen to have gone up much more than the costs: from a mean price of 18 kr/kg in 2003 to 37 kr/kg in 2010. Because of this, the finance director explains, they are actually doing OK.

The FCR helps us understand how an entity like salmon can inhabit multiple realities at once: one submerged in water, another in boardrooms and at Monday morning meetings. Out in the fjord, the FCR (especially the biological FCR) may be one of many concerns that inform day-to-day decisions concerning the treatment of each cage or pen. Here, it can perform a function similar to that of "checking on the feeding," which was described in chapter 3. But most importantly, the FCR (both types) is a quantitative measurement of relative performance at each production unit, which can be used in evaluation exercises at the head office and beyond and can also guide investments and shareholder decisions.

When we consider these two salmon figurations—one a living, growing entity under the water, the other an aggregated figure on an electronic screen—it is not as if one is more real than the other, or as if the latter is an abstraction of the former. They are both real, and they are both material too. Both are dynamic entities in the processes of domestication, processes that involve widely different heterogeneous practices. Drawing on Annemarie Mol's work (2002), we could refer to these entities as "salmon multiple." These are not radically different worlds but carefully crafted outcomes of specific forms of world-making. What is at stake is the relation between the two and the way in which they work together to render salmon as biocapital. I suggest that this is how the FCR becomes particularly important. Mediating across sites and latitudes and between salmon fleshiness and electronically submitted numbers, the FCR performs the role of what Star and Griesemer (1989) identified as a "boundary object." In a later publication, Bowker and Star (1999) referred to boundary objects as "plastic enough to adapt to local needs and constraints, yet robust enough to maintain a common identity across sites" (297). As such, boundary objects contribute to "developing and maintain[ing] coherence across intersecting communities" (297).

An FCR mediates between biology and economics, but its properties as boundary object facilitate other mediations as well. Because a low FCR indicates a more cost-efficient production, the uses of FCR render its units comparable along a normative scale where some are better than others (that is, both better at what they do and potentially more profitable). Unlike simpler measurements (such as temperatures or quantities of fish), the FCR both enacts and can be used to reinforce a normative order.

An FCR can be performed for a single tank or pen, for a production site, or for a firm; and it can be used to compare the metabolic efficiency for different species (for example, salmon and rainbow trout, or chicken and pigs). It can also be calculated as an average figure for an entire region or even a country. Elsewhere, I have described how the FCR allows comparison across sites, and hence is integral to universalizing salmon aquaculture as a field of expertise. Through such comparison, Tasmanian salmon production can be displayed as "less efficient" than Norwegian production, which often figured in Tasmanian business meetings as the standard that they all should aim for (for details, see Lien 2007, 178–80). Because biomass is central to any calculation of FCR, it is essential for evaluating salmon aquaculture as a profitable enterprise, as well as for calculating the amount of salmon available for slaughter (see the section "From Biomass to Edible: Logistics and Export" later in this chapter). In this way, biomass is key to calculations related to salmon farming as a business enterprise.

Biomass mediates other relations as well. Until 2005, Norwegian salmon production was regulated by so-called feed quota, which implied that public authorities placed an upper limit on the amount of feed available for salmon farming for each company, or license.[24] In 2005, this was replaced by a new set of regulations that defined maximum allowable biomass (MAB; *maksimum tillatt biomasse [MTB]*) as the regulatory instrument to control the growth of the salmon industry. These regulations require reports from each locality and company regarding the amount of salmon biomass per unit (cages or pens) at the end of each month. *Biomass* is defined as live fish, measured in kilos,[25] and it is reported electronically via a platform called Altinn.[26] Hence, once the operation managers have submitted monthly reports for each locality (and after internal quality control), the numbers are entered into a statistical database, where they are then aggregated and broken down by county and published on the Directorate of Fisheries website. In this way, our daily counts of morts become, eventually, part of the national statistics that tell how much farmed salmon is being raised in various parts of Norway each month. This constitutes, in turn, an instrument for policy evaluations and regulations at the national and regional levels.

BECOMING BIOMASS: PRACTICES OF WEIGHING AND COUNTING

Biomass and FCR are key components in operation managers' reports to their head offices. They are done for each tank (monthly for smolt production) and for each cage or pen (weekly for grow-out sites). But how are they done?

"We make your fish talk!" was the slogan of a leading supplier of automatic feeding systems, Akvasmart (which later merged and became Akvagroup), which won an innovation prize in 2005 for its cutting-edge software, Fishtalk.[27] At Sjølaks, Fishtalk is a key tool in reporting, particularly in relation to FCR, because it performs calculations automatically, based on input related to feed (sometimes monitored by automatic feeders) and biomass statistics. When Tore reports to the head office he uses Fishtalk. Let us use his September report as an example. On a matrix, in which the font has been made very small so that it fits into one sheet of paper, I count fifteen rows, one for each tank (plus a sum total), and no less than twenty-three columns for different numbers, all of which describe various measurements related entirely to fish and feed. Let us take a look at what some of the columns consist of.

Table 1 is a simplified version of a monthly report and shows figures Tore reported for tanks 11, 13, and 14 in September. To understand the figures and the reality they enact, it helps to imagine the format of a financial report with its incoming and outgoing "stock." The columns labeled "Incoming" indicate biomass at the beginning of the month, and the columns labeled "Outgoing" indicate biomass at the end of the month. Biomass is calculated as the number of fish times the average weight for each tank, and thus enacts salmon as a collective, identified by a specific tank or cage number and locality. This entity, which I refer to here as a "batch," figures as the smallest economic and biological unit in the production cycle. This is how individual salmon are traced, known, and identified.

The difference between the two points emerges as the outcome of two biological processes: growth and death. We know from chapter 3 that *daufisk*—dead fish, or morts—are counted as they are removed, the number carefully reported and added to day by day. The 376 dead fish found in tank 13 (see table 1) during the month need to be subtracted from the incoming number; there are slightly fewer fish left in the tank at the end of the month. But even so, the total biomass has increased, from 3,387 kg to 3,997 kg. This is because each fish has grown bigger. But how is this known? To record the weight of all salmon in an entire tank or pen is practically impossible. Their aggregate weight can be revealed only upon slaughter, when it is technically possible (economically feasible and ethically justified) to weigh each fish one by one. Before that, salmon biomass is calculated as an estimate based on sampling.

Sampling procedures differ during the salmon life cycle. It all begins at the time of first feeding, when fragile alevins (having consumed their yolk sac) are moved from trays into tanks. Not only is this the moment of transitioning from feeding off their yolk sac to being fed by us, but it is also the first time they are enacted numerically as a collective, as a batch. Before this, they were layers of orange eggs, known as weight but never counted. Now, for the first time, every single alevin is enacted as part of a larger aggregate unit. Hence, from now on they are enrolled in the calculative apparatus of the firm as an entity that is legible, economically as well as biologically. In chapter 5, I will describe this procedure in greater detail. For now, we simply need to note that each stage of the salmon life cycle is associated with specific challenges regarding the enactment of salmon as biomass.

Counting alevins, as we shall see, requires huge approximations, so the numbers produced are somewhat uncertain. At the smolt production site, it is different.

TABLE 1. Excerpt from monthly report, Frøystad smolt production site, September 2009

Tank #	Incoming				Outgoing					
	Number of fish	Mean weight (g)	Biomass (kg)	Mortality number	Number of fish	Mean weight (g)	Biomass (kg)	Daily growth (%)	Feed fed (kg)	Economic Feed Conversion Rate
0011	63,813	48.0	3,055	1,016	62,797	58.9	3,698	0.99	763	1.19
0013	54,752	61.9	3,387	376	54,376	73.1	3,977	0.80	676	1.15
0014	55,969	59.3	3,316	292	55,677	67.0	3,731	0.59	626	1.51

SOURCE: Reprinted with permission from Sjølaks (pseudonym).

No longer fry (they are now referred to as *småfisk*, or "parr") and with an average weight of about 50 grams, they can easily be held in the palm of the hand. That means that they are also big enough to be sluiced through an automatic sorting device that counts them individually.[28] In this way, the batch can be recalibrated and numbers provided like the ones in the first column in table 1. Once you know the number of fish, all you need in the following months is the number of morts, and the current number of fish can be adjusted accordingly. Average weight, on the other hand, requires regular sampling. To know that the fish in tank 13 are bigger than the ones in tank 11, or that they have grown larger during the previous month, you need to take a sample of one hundred fish and weigh them individually or, more typically, in a convenient number—say, twenty at a time, using a bucket full of water. With fish between 10 and 50 grams, this task is not so difficult. But even though sorting machines are helpful, they are not always reliable, as my fieldnotes show:

> "In my next life I want to produce nails!"
>
> It is early July 2011 and I have just run into Tore down by the harbor. I have spent most of my time out on the fjord this summer and haven't seen him for a few months. His exclamation is followed by a sigh, but as usual there is a smile curling up the corner of his mouth. I ask him what is going on now. He explains that they have had fresh delivery of parr and have run them through the *sorteringsmaskin*, the automatic sorting device, which can be preset allowing different sizes to escape through to different pipes, leading to different tanks. But the sorting has revealed some problems.

Individual fish grow at different speeds; hence, a fish delivery will always contain fish of different sizes. But variation within each tank should be avoided. It is assumed that the big fish will dominate and eat more as a result and become even bigger, while the small become even smaller and more vulnerable to disease. Furthermore, a fairly uniform size among the fish in each tank is likely to ensure more appropriate collective treatment of the batch as a whole.

Tore complains that the problem with the most recent delivery is that too many fish were too small. Tore reckons that about half (rather than the expected third) were in the "small" category, which meant they went straight through the roster, while only one in six were sorted as "large." Consequently, and in order to fill up the tanks with equal numbers of fish, he had to reset the sorting machine, run the entire group of "small" once more, and then redistribute the whole lot into three tanks with smaller-than-usual versions of the categories of small, medium, and large. This means he will now be about 30,000 short on his planned September delivery of 350,000 smolts. And there will be delays in vaccination too: they have to weigh at least 10 grams before he can vaccinate. Many of them do not. I ask:

"Can't you just order another delivery from the supplier?"

"That won't do. The fish are already into the winter regime,[29] and a new delivery would be out of synch with the rest."

In spite of clever regimes to produce fish as standardized, uniform batches, the fish will always resist our efforts. Some of them deviate from the norm and grow at a different pace. Tore is probably thinking along those lines too when he jokingly announces his plans to produce nails rather than fish in his next life.

Out on the fjord, the practices that feed into FCR calculations are different again. From their delivery as smolts until they are sent off to slaughter some eighteen months later, the fish will increase their weight by a factor of one hundred (from approximately 50 grams to 5 kilograms). This period, the grow-out phase, is all about putting on weight. As the salmon grow, the feed they consume increases dramatically as well, and thus feed accounts for an ever greater portion of the total monthly cost of production. This means that keeping an eye on the FCR is more important than ever. But it is tricky. A salmon at 3–4 kg can be lifted up with a dip net and weighed individually but not very accurately: its liveliness and its obvious discomfort out of water disturb the finely tuned calibration of the electronic scales. In order to avoid such handling, a "biomass-measuring device" *(biomassemåler)* is used. This is a kind of frame that is lowered into the pen. After a while, fish begin to swim through it, more or less randomly. The frame has an electronic sensor that records their volume and calculates their weight (that is, their biomass) automatically as they pass. It can be programmed for a specific number of entries—for example, one hundred (i.e., the first hundred that enter)—and then it will provide their average weight. But this is not accurate either.

An operation manager explained the difficulties in an April 2014 e-mail (my translation): "As always, there are sources of error. Large fish tend to swim deeper than the small ones, who tend to swim along the edges of the netting. So if you place the frame deep down, you will have another average weight than if you place it further up, and so on."[30] "At the end of the day," he continued, "estimating biomass is an ongoing process in which you consider the number of fish placed in each pen, the amount of feed that has been distributed, and also the average weight as expressed by the biomass measuring device." The absolute biomass value is only revealed after slaughter. This is when operation managers have a chance to learn how far off the mark their estimates have been: "While we usually hit fairly accurately," he reported, "there is a tendency for most operation managers to estimate a higher biomass figure in each pen than there really is. As a result, their calculated FCR at the time of slaughter can be a source of disappointment."[31] His comment indicates that biomass estimates are based on a kind of tinkering, which is not that different from what goes on when checking on the feeding. Estimating biomass is therefore not so much a matter of producing perfect and precise data—approximations are unavoidable—but rather a matter of rendering salmon visible as numbers that can then be manipulated in various ways.

"TRUE COLORS": SAMPLES AND ADJUSTMENTS

During the early stages of fieldwork, I was struck by the way in which smolt production sites, and to a lesser degree also the grow-out sites, were also sites of knowledge production. Surprised by the vast amount of numbers produced at each locality, I began to think of farmhands as laboratory workers in disguise, hard-pressed to allocate sufficient time to a challenging coproduction of numerical data and salmon flesh, or numbers and smolts. With John Law, I toyed with the image of a salmon farm as a center of calculation (cf. Latour 1987), each locality supplying numbers that together coconstituted a dynamic distribution system of knowledge central to contemporary marine domestication. This is not entirely wrong, but it overlooks the friction that occurs during translations from lively flesh to numerical figures. It overlooks how flesh and numbers are always only partially connected and how they sometimes resist being brought into precise alignment.

The following fieldnotes were mostly put together one evening after John and I had shared a dinner of fresh salmon. We are back at Vidarøy in June 2009, when an earlier batch of salmon was scheduled for slaughter. The salmon figures in the fieldnotes too. It had to be killed anyway and was divided into serving portions on the platform. We were consuming our field material as we wrote, so to speak, and it tasted all right.

> Fredrik explained about the color sampling a few days ago. If the salmon flesh is not red enough, they may switch to a different type of feed. Fredrik shows me at the computer how he logs into Biomar (current feed supplier) with the username "Nils" (the previous operation manager) and then he scrolls down a list of different types of feed. Right now they are feeding a type which has the suffix "ss20." It could be ss10, or ss40, indicating more or less color. "They" will decide on the feed to order next, he explains, and nods in direction of the shore, where the head office is located. "They" also decide the exact time of slaughter. According to the current plan, all fish at Vidarøy will be slaughtered by October, but the exact schedule of slaughter is constantly adjusted according to changes in market prices as well as in the pens, and also in relation to the capacity available at the Sjølaks slaughtering plant.
>
> Today is the day of color sampling. I ask if I can help. Karl says I can help write down the numbers while he lifts the fish out of cage 4.[32] All the equipment is out on the platform already: a table, a net, a scale, a small wooden ramp, a cutting board, a prepared form with two columns for weight and length, and five rows for the number of fish for each cage, small pieces of paper for inscriptions like "cage 4, fish #2," a sharp knife, a bucket of water, an empty bucket, a ruler, a few plastic bags, and a

pencil. I later learn that the bags will go straight into the freezer and then be sent off to a laboratory elsewhere, where the amount of color in the flesh will be measured chemically. Based on these results, it will be decided (at the head office) whether to continue feeding as before, or increase or decrease the amount of color in the feed.

In the top row of the column indicating weight (grams), next to the cage number, I notice that someone has already written "4560." Karl explains that this is the average weight of fish in cage 4, based on current biomass estimates. He throws a few pellets on the surface and soon he scoops up a fish, and with some effort, with the handle of the dip net resting on the railing, he lifts a wiggling salmon up above the water and then onto the platform, where he immediately kicks its head. A few convulsions and it is dead, still on the cutting board, and we can place it on the scales.

We both look at the scales and watch the digital numbers settle at 5400, whereupon Karl says, "Five kilos" and places the fish on the wooden ramp. He quickly gets the knife out and cuts it into three pieces. Only the middle section is used for color sampling. With some hesitation I am about to write "5000 g" in the column made for fish number 1 in cage number 4, but then I ask: "Are you sure?" Karl nods, and without any further explanation he shoves the front and back end of the fish into the empty bucket, rinses the three middle sections in the bucket of water, then hands them to me. I stick small pieces of paper with the inscription "cage 4 fish 1" on the skin of each piece, place each piece in a separate plastic bag, and tie a knot at the end of each bag to seal it. In the meantime, Karl is already bending over the railing, trying to catch another one. The next fish is much smaller, the scales stop at 4.800, and Karl says the number aloud exactly as I also see it. I write "4800" in the row marked fish 2. And so on. We continue until we reach the number 5 and then move on to the next cage.

I ask whether these are good to eat, and Karl says that these are perfectly fine salmon. He cuts a few fillets from a leftover tail, and suggests that I put it in the plastic bag and take it home for dinner. Sometimes several fish end up in the dip net, sometimes none. Karl needs to catch only one at the time, and they are often difficult to catch. After a while I ask:

"Why did we write 5000 grams for the first fish when the scale said 5400?"

Karl hesitates, then explains in a way that makes me understand that he knows that what he did was not exactly according to the book: "Oh, you see, it has to be fairly close to the average weight. The first one I caught was a very big one. You are allowed to deviate a few hundred grams, but not too much."

The "it" refers to the number inscribed in the box for each fish. "Average weight" is the estimate inscribed on top of each column, and it refers to the cage as a whole. Karl explains that they would normally use Benzoak (an anesthetic) and then weigh several fish to select the ones that are closest to the average weight and within the prescribed size range. That way, they would not have to kill the fish; they could all go back into the cage after being measured, and if they caught a few too many, it would not matter. But since Benzoak has a retention time of twenty-eight days, and because we cannot know for sure when this cage will be slaughtered, we cannot use it now. He doesn't say much more, but what I gather is this: In order to weigh the fish, it must be dead, because when it is alive, it is very lively and slippery. It is rather difficult to judge the size of a fish before you actually catch it with the dip net and lift it up. So what are the options available? If Karl writes down the accurate weight of a fish that exceeds the accepted deviation of about 10 percent from average weight, the sample will not count as representative of the cage, and the number might be discarded down the line. In other words, , the piece of flesh will be useless as data for judging the true color of fish in cage 4, and our sample will then be smaller then what is required. (I am assuming that scientists in the lab will act in accordance with protocols that state that the variation cannot exceed 10 percent of the average size, based on knowledge that difference in color correlates with difference in size.) Knowing that this is what would have happened (that the inscription would be worthless anyway), Karl might have chosen to throw the fish away and catch another one, to reach the sample size of five. But if he discards the flesh from a nice-looking, healthy 5.4 kg salmon that he has already killed, it represents an economic loss and means that the salmon was killed for no reason. Benzoak would have let us anesthetize the fish instead of killing it to get its accurate weight, but then, slaughtering would have had to be postponed for at least four more weeks, and they would have fewer batches available for slaughter in July. Faced with this dilemma, Karl stretches the rules where it is likely to do least harm. He violates the principles of random sampling and standard deviation for judging a shade of red, which is, after all, hardly consequential for anyone. Through this choice, Karl reveals himself as a less-than-accurate laboratory assistant but a very sensible farmhand. Producing numbers is important, but what is more important, after all, is producing fish for slaughter.

Unlike in scientific practices, numbers are not produced primarily to verify or falsify hypotheses or to substantiate propositions about "the real." The point is not that they cannot make such propositions—obviously they can—and numbers are often mobilized among farmworkers to suggest causal connections of various kinds. But data perform other kinds of work as well, which are more committed to the ongoing politics and practices of salmon farming than to generalizing statements about the nature of biological entities. Most important, the numbers we produce mediate across the various organizational layers of the company, as well as beyond, to regulatory authorities, exporters, and society at large. Often, the whole purpose of assembling numbers is to get other things done. Figuring as what actor-network theory identifies as "obligatory passage points" (see Latour 1987), a weekly report or a smolt delivery form helps enact the transition (of smolts or of budgets) from one phase to the next or one location to another, and thus it simultaneously enacts and aligns the various temporalities of the domus assemblage.

FIGURE 5. Salmon samples for color analysis (photo by John Law)

When we have finished color sampling, I walk past Kristoffer, the head operation manager, and ask about the latest plans for slaughter. He says that cage 1 will probably "go on hunger" (*gå på svelt*). In other words, they will stop feeding, in order to ensure that the guts are empty. (This is common practice during the last two weeks prior to slaughter.) But it is not quite confirmed yet, and he adds: "We will have to see what the 'chief' says" *(Me får sjå ka høvdingen seie)*.

Half an hour later he is on the phone with the head office. Finally it is confirmed: Cage 1 *will* be placed on hunger tomorrow but fed to the end of the day today. Cage 2 will be placed on hunger the day after tomorrow. But what has happened in the meantime? How is this decided?

So far in this chapter, I have explored various ways in which biomass numbers mediate between localities and the head office as well as between the firm and regulatory authorities. But biomass numbers also provide a dynamic estimate of the amount of salmon available for slaughter at any given time. As such, biomass can be seen as a measure of inventory, or potential biocapital. It is not capital that can be reinvested quite yet: while the fish are in the sea, anything can happen. But provided that things go well, biomass numbers offer an indication of the amount of salmon that can be "activated" as capital for further investments, through the process of slaughter. In this way, biomass figures are crucial in the strategic calculations about when to slaughter, decisions that consider not only what is available under the surface but also the expected fluctuations of global spot prices and delivery requests from Norwegian exporters. To conclude this chapter, let us turn to the head office, where, figuratively as well as literally, these numbers are tied together.

NOTES FROM *HI SIO*: CALCULATIONS AND SALES

I am spending a week in the main office. Two years after our first field visits, I decide it is about time to get a grasp of salmon from "the other side"—from *hi sio*. After John leaves with the crew for the grow-out site every morning at 7:00 and my daughter, Eira, who has come with us this week, catches the school bus at 7:30, I still have time for another cup of tea before *my* field site wakes up, leisurely, between 8:00 and 9:00.

> The contrast between the offices and the localities elsewhere is striking. Entering through the main door I feel as if I am stepping into a hotel lobby. Sheltered from the wind and rain, the large reception area is decorated with modern art and surrounded by small cubicle-like offices, with glass doors that offer solitude for people and their computers. The common room has a rustic lunch table of heavy oak; cool, elegant kitchen areas; dried apricots, nuts, and chocolate; coffee and tea. The building is spacious, light, all wood and Scandinavian design, and appears to have been built by a company that is optimistic about its future.
>
> It is my third day in the office, and I am beginning to feel as if I am wasting my time. People are busy. Or they are not here. Interview appointments are being rescheduled. The soft buzz from Monday morning has been replaced by a nearly complete silence, only interrupted occasionally by the female receptionist answering the phone upstairs or the even, subtle rhythm of a copy machine that is busy producing copies, very many copies. It is the familiar incarnation of work in our day and age: "the office," where everyone is busy doing something out of sight and only barely heard—the subdued voices of phone calls behind closed doors, the tapping of fingers on keyboards in cubicles whose doors are open.
>
> Is salmon here as well? I glance through John's fieldnotes from the day before. Our workdays are completely different. No wonder my coworkers out there wonder about what office workers do all day when over there, among the tanks or pens, there is carrying, emptying, feeding, measuring, feeding again, keeping alive, caring for, attending to, watching, making sure the gloves fit tight and are warm enough, stretching hands down under the water, grabbing, holding on to, losing, and letting go.[33] I am thinking that some of what is going on there is going on here too, but how? What is the *materiality* of these practices here in the office? At the

moment I cannot think of a better question to ask than this: How can salmon be done in a place which is so *dry?* and so comfortably *warm?* where our bodies are so comfortably cared for, where the chairs are soft and reclining, and the coffee is always warm and always freshly made?

From the door of my cubicle I see next to nothing. So I decide to follow the numbers. And I engage office workers through the format of an interview. It is not perfect, but it is better than nothing. And it soon turns out that even if numbers have different destinations within the head office, nearly all of them eventually end up at the table of the director of finance:

> Håkon grew up on a nearby island. Like the owner, he is considered local by most people, but localness in this region is always specific. A subtle difference in the way he speaks reveals that his formative years were spent on an island some 50 km further north. Now we are sitting in his office, and I have just started my tape recorder.
>
> *Marianne:* So let us start with Monday mornings. What sorts of things have arrived by then? What is there for you to look at?
>
> *Håkon:* Well, my papers, all these files and forms—they end up in a stack right here. (*Sound of paper being lifted.*) You have probably seen these already, haven't you?
>
> *Marianne:* Yes, I am sure I have seen many of them already.
>
> *Håkon:* So you see, this is . . . this is a lot . . . but the most important thing for the Monday meetings is to go through production, especially at sea, at the grow-out sites. You see, lots of forms. Let us look at the main one, if I can find it . . . Here!

As we are speaking, Håkon immediately lifts up a pile of papers, the incoming numbers for last week. They are stacked on the left side of a large desk, which was already crowded with other documents. As we speak, Håkon describes his work in a way that indicates that it has more to do with assembling numbers in different ways, grouping them differently, comparing them across time and space, and against external thresholds (of which the maximum allowable biomass [MAB/MTB] is an important one) than with papers as such. The numbers provide a lot of details for the managing directors to consider, and Håkon sees this as a competitive advantage:

> *Håkon:* Many of us [in the office] even if we are not directly involved in raising fish, we know a lot about what is going on out there. Compared to other firms, I dare say that our management is extremely focused on practical management [Norwegian: *drift*]. You will hardly find a salmon firm in which the owners and managers are so hands-on and

> are so dedicated to practical details. I think that is why we are doing so well.
>
> Let me illustrate: Today they are submitting production figures. And it will only take you a minute, and you will see it, for example, if there is increased mortality at a particular site. What is important is that if something like that happens, we need to comment upon it because you may only have a few days to turn around, to do what needs to be done.

Although salmon manifest here as an unruly stack of paper, about 5 centimeters thick, casually picked up by Håkon at the beginning of our interview, this is not all they are. It is not even the main thing they are. Had they existed as paper forms only, I think Håkon would have been more careful and archived them in a more orderly manner. The papers are simply printouts from a database, practical worksheets, in which it is the relation between specific numbers that really counts. During our interview, it becomes clear that even though most numbers pass through Håkon, the responsibility for these data rests with someone else. I am directed to Finn, and schedule another interview for the following day.

Finn spends about a day every month preparing the required monthly report on biomass to the authorities via Altinn. Then he spends every Monday from early morning until noon looking at numbers dealing with production at various localities. Finn is also responsible for internal quality control of all forms submitted from grow-out sites and does some work for the slaughtery and smolt production sites as well. According to Håkon, he has figures at his fingertips. So I ask him during an interview the following day, "What kinds of 'paperwork' are circulated within and outside of the company? How can we get an overview?"[34]

Håkon had pointed to a stack of paper on his desk, but Finn immediately starts sorting paperwork according to temporal sequences rather than by content. The temporal, cyclical rhythm is important; it structures his week. In my own fieldnote folder, I have archived many of these forms as printouts from various sites, but he sorts them differently, in a way that cuts across localities. Detailing a cyclical sequence of reporting events, he draws a picture of the firm (or head office) as a node in a network that reaches far beyond the firm itself. Four main categories can be identified:

1. *Vekeskjema:* weekly reports from the localities to the head office. This is what is entered through Fishtalk. For grow-out sites, the data reported includes feeding, morts, mean temperatures, lice counts, lice treatments (if any), and sometimes salinity and water transparency. They are entered by the operation managers at the end of the week and are then immediately available at head office. Before the Monday morning meetings, figures are summarized and checked by Finn for a general overview of the current

state of affairs. (Smolt production sites produce a similar report every month.)
2. *Månedsrapport* via Altinn: monthly report from the firm to regulatory authorities. On the seventh of each month, each license-holding company submits a report through Altinn.[35] While the report is the same, there are two separate recipients: the Directorate of Fisheries and the Food Safety Authorities. For grow-out sites, key parameters are number of fish, number of morts, mean weight, and biomass both at the beginning and at the end of the previous month; feed distributed; slaughters (number and biomass); escapees (if any); lice counts; and lice treatments (if any). This is the basis for the implementation of MTB regulations.[36]
3. *Vekerapport til Fiskhelsenettverk*: weekly report to the Hardanger Fish Health Network (Hardanger Fiskehelse Nettverk; see also chapter 3, "The Salmon Domus as Laboratory: Becoming Scientific Data"). As part of regional measures to mitigate sea lice (implemented to protect migrating wild salmon smolts), every salmon license holder is required to submit weekly reports on the prevalence of sea lice. Reports, as well as mitigation measures, are organized to cluster companies according to their location in relation to smolt migratory routes. Vidarøy and ten to twelve other producers in the area are required to share information with one another. The Fish Health Network provides a standardized form that is used to coordinate treatments at a regional level. The forms are prepared weekly by each operation and submitted to the head office, and Finn passes them on to the FHN secretary.
4. *Månedsrapport* to Sjølaks: In addition to the above, each locality produces a separate monthly report to the head office, providing detailed comments not only about current figures but also about personnel, production outcomes, challenges, and other matters. This is an opportunity to report about things that are not covered by other figures and is for internal use only.

So far, we have focused primarily on figures as they are sluiced, sorted, and corrected by the head office, a center of calculation for the firm as an economic unit as well as an obligatory passage point that connects each locality to external networks and regulatory authorities. But how are these figures mobilized in relation to decisions about slaughter? What had been going on before Kristoffer, in the early afternoon, received a phone call from the head office instructing him to place cage number 1 on hunger?

Slaughter is organized according to a detailed schedule *(slakteplan)* that is produced many months in advance of actual slaughter. This gives operation managers, as well as the slaughterhouse, well boats, and the head office, an idea of approximately when different localities will provide fish for slaughter. But the slaughter plan is only tentative. The final decisions are made by the managing directors, and their various options are evaluated continuously. Håkon explains: "Decisions about when to slaughter reflect overarching concerns, including the budget. They emerge from our attempt to utilize our MTB (maximum allowable biomass) as accurately as possible. We also try to "hit the market" *(treffe markedet)*. And we try to slaughter when profit is at its peak. In addition to this, there are biological concerns; for example, if there appears to be an emergent sea-lice outbreak, then it is better to slaughter sooner than later."

FROM BIOMASS TO EDIBLE: LOGISTICS AND EXPORT

The roads are icy. It is a dark Monday morning in December, and the trailer truck has parked near the shore, next to a brand-new salmon slaughterhouse and processing plant that forms the backbone of the industry in this small coastal community. Inside the plant, the workday is about to begin. Within the next six to eight hours, the forklifts will drive back and forth, moving stacks of Styrofoam boxes full of salmon from the processing plant to the temperature-controlled storage room inside the trailer, while the driver takes a rest. The salmon were slaughtered yesterday and will be processed today. Salmon scheduled for slaughter tomorrow have already reached their destination and are stored in a pen close to shore. Tomorrow morning they will be sluiced into the slaughterhouse, through a huge transparent plastic pipe. If everything goes according to plan, 20 metric tons of freshly packed salmon will fill the trailer by 3:00 in the afternoon. The driver will drive directly to Stavanger, catch a ferry to Denmark, and drive from there to Schiphol, near Amsterdam, where there are plenty of passenger flights with capacity for extra cargo, and cargo flights as well. The trip to Schiphol takes about thirty hours. At Schiphol, the boxes will be stored overnight and then transferred to flights with Asian destinations: Tokyo, Taipei, Seoul, Singapore, Bangkok, or Dubai. In the meantime, other trailers will arrive and depart for Paris, Brussels, or Berlin, and salmon from Vidarøy will be distributed from them to local seafood auctions or markets. By Thursday, the salmon will be served in restaurants or picked up in supermarkets in cities practically all over the world.[37]

This is just a glimpse of a sophisticated network of transport and logistics that involves some ten to twelve Norwegian exporters, many more salmon producers, and an infinite number of buyers, including importers, restaurants, and retailers. But how does this sudden realization of salmon as biocapital work?

Basically, everything is settled in a few hours every Friday morning, as briefly described at the beginning of this chapter. The temporality of export is a weekly cycle. By Thursday night, the Sjølaks sales director has studied the biomass figures from various localities and has a fairly good sense of the amount of salmon available for slaughter, their sizes, and the urgency of getting them out of the water. He may seem busy on the phone all morning, but the real centers of activity are elsewhere, in Ålesund, Bergen, or Tromsø, where the exporters' offices are located. Here, a cacophony of different languages fills the spacious office where each employee occupies a small cubicle. Switching back and forth between importers (in Japanese, Chinese, Spanish, French, English, and other languages) and producers (speaking mostly Norwegian), they negotiate prices, shipments, and sizes until they strike a deal, which means, in practice, a contract promising salmon of a particular size at a particular time delivered at a specific port or airport at a specific negotiated ("quoted") price. Telephones are important, but Skype is useful too.[38] While a medium-sized company like Sjølaks may deliver to only a few exporters,[39] each exporter usually negotiates deliveries from many producers and has even more importers on its contact lists. With each shipment, the exporter will make a marginal gain, calculated as the difference between the price quoted by foreign customers and the price accepted by the supplier (minus the cost of transport and other expenses). Based on the prices offered by a particular importer, the exporting company may offer a bid to companies like Sjølaks. A bid may stand for an hour or for only ten minutes, depending on the situation. But because Asia is an important market, most deals have to be made by 1:00 p.m. in Europe in order to be finalized before Friday evening in Asia. In this way the negotiations serve as a kind of auction, but one in which exporter and supplier are mutually dependent: Sjølaks needs to sell, but the exporters also need a trustworthy supplier in order to strike a deal with their customers, because they cannot take any orders if they cannot deliver.

This means that when the salmon are prepared for the journey from their grow-out site to the slaughter (see chapter 6 for details), their final destination is already settled. They will already have been "sold" at a specific price, and their final journey overseas will have been planned. Somewhere, in a fax machine or in an e-mail inbox, their *value* is quoted as a number—for example, 36 kroner per kilogram. Hence, a fish of 5 kilos represents 180 kroner of fleshy, lively salmon, provided that its skin is flawless and that future handling is done with care (if not, it may drop from "Superior" to "Ordinary" and fetch a lower price). Now and then such specific constellations of monetary value will trigger further calculations about profit margins for specific localities and for the firm as a whole, as well as about feed efficiency and FCR. But such calculations are mostly confined within the walls of the head office. Out at Vidarøy, the salmon, which are now *on svelt* ("on hunger"), may begin to wonder where the pellets are. Soon they will be relocated from what has been their home so far and experience the inside of a well-boat tank (see chapter 6). They are still alive, but for most calculative purposes, they are already on their way to becoming a commodity. As biocapital, their destiny is sealed.

While they are still in the sea, or in the processing plant, the salmon are technically the property of Sjølaks. But as soon as they are placed in the trailer outside, they are the property of the exporter, who remains the owner until the shipment is handed over at an airport or a sales destination elsewhere. And this is, basically, the end of the story as far as Sjølaks is concerned. Even if markets are a huge concern, they materialize as fluctuating prices from one Friday to the next. Other than that, the directors at Sjølaks do not know very much about where the salmon ends up, who eats it, and how it is consumed. Nor do they worry about such things. As long as global spot market prices are reasonably high (and if they are not, there is not that much they can do about it), then their main concern is to deliver sufficient quantities of sufficiently high-quality salmon biomass. This requires a great deal of knowledge, persistence, and effort over several years and a keen sense of business and investments as well as the practical operational management of places like Frøystad, Vidarøy, and Idunvik. But it does not require that you know a whole lot about salmon consumers.

SHIFTING CONFIGURATIONS

In this chapter we have traced the transformation from farmed fleshy salmon to biomass and seen how "salmon-as-biomass" serves as a boundary object, mediating between different localities, and facilitates mobility along different trajectories. While fleshy salmon are penned up until they are carefully hauled to the slaughtery and become globally mobile only shortly after processing, salmon as biomass can appear in many locations at once, unconstrained by the logistics of transport and linear time (and hence the temporality of flesh disintegration and decay). I have suggested that rather than seeing one as real and the other as abstract, both emerge as material entities, carefully crafted and specific forms of world-making. Both are designed to perform particular tasks in the production of biocapital, and both make a difference in the world. And yet, they are not identical; their overlap is only partial. Although they may sometimes appear as seamless reflections of one another (the biomass *representing* the salmon directly), they are also sometimes incoherent and there is friction between the two.

Their difference appears most clearly in situations when one is seen in light of the other. This does not happen all that often. In the head office, for example, and for the sales director, it is possible to engage in calculations about biomass and delivery dates without really considering the salmon as fleshy beings. The sales director needs to assume that the figures on his computer reflect a certain amount of salmon available for slaughter. He needs to trust the numbers, and unless something unexpected happens, he can engage with numbers as *real*. Salmon *are* figures on a screen.

Out at Vidarøy, it is a little different. The presence of the cages and the constant tinkering in relation to the fishes' needs make the fleshy and lively reality of salmon an ever-present "real." As Fredrik, Karl, and Aanund enter the figures on their computers, they are painfully aware of the uncertainty that is inevitably attached to these numbers from time to time. Constantly juxtaposed as figures and as flesh, the salmon at Vidarøy remain multiple. But they need to be somewhat coherent. No wonder, then, that it is possible for a new operation manager to mess it up, such as the unlucky operation manager who saw his numbers thrown into the wastebasket at a Monday morning meeting. His mistake was not that he did not care for salmon but rather that he forgot to pay due attention to the fact that an operation manager does not produce only salmon. He produces numbers too. And both should be of good quality, always.

Hence, we could say that as salmon travel as biomass from one site to the other, their move is a shifting journey from many to one. We could describe this as a familiar route of commoditization: biomass allows the calculations that make salmon a fungible commodity on spot markets. But that is not all there is. The route towards the spot market is only one journey that domesticated salmon can take in their trajectories of becoming. As we shall see in the following chapters, salmon are about to become other things too: scalable, sentient, and even alien. We will elaborate on each of these in due course. But first, let us turn to industrial expansion, which is the topic of the next chapter.

5

BECOMING SCALABLE

Speed, Feed, and Temporal Alignments

If domestication is a relation, how can we describe this relation in salmon farming? What is it made of? What kinds of textures are involved? Marilyn Strathern (2006) reminds us that relations can be many different things: connections, affinities, causes, effects, or particular forms of storytelling. Narratives of domestication foreground the human-animal relation as one of confinement and control. In previous chapters I have told the relation differently, emphasizing elusiveness and uncertainty as two of its key dimensions. I have highlighted biomass and numerical calculations as tools that make the salmon assemblage manageable. In this chapter I dwell on similar sets of relations but from a different perspective. Here my aim is to try to understand salmon farming not only as a story of how salmon become domesticated but also as a story of industrial growth.

When Norwegian pioneers shifted their attention from hatcheries to marine pens and began to breed Atlantic salmon through their entire life cycle, it was indeed a significant move. It was a further tightening of relations within the salmon assemblage and can be seen as another turn in the human history of domestication (see chapter 3). But in itself this turn tells little about profitability, scale, and economic growth. Imagine salmon farming in the 1970s: a small-scale operation, scattered along the coast, ensuring no more than a steady supply of fresh salmon to local communities. It could have stayed that way. But as it turned out, a different scenario unfolded, one of rapid expansion, exponential growth of "biomass output," huge financial investments, and successful export. The last few decades of Norwegian salmon farming have involved unprecedented growth, both in terms of export revenue and in terms of biomass.

This chapter explores the textures and temporalities of this expansion. It is about entities that keep on multiplying and the continual creation of a new normal. This was most evident at the hatchery and smolt production site: "If you currently produce a quarter of a million smolts each year, then why not double that?" "If you thought 10 meters across was a big tank, then take a look at our new one. It's 16!" "If you thought your current system was efficient enough, think again." I am interested in the kind of growth that takes place while the basic elements of the salmon assemblage stay more or less the same. I am interested in growth as multiplication and how it is done.

Salmon aquaculture is among the most expansive fields in global food production, but there is an important caveat: unlike most other expansive food industries, economic growth in salmon aquaculture does not rely primarily on value adding. "Value adding" refers to the ways in which foods are transformed so that they can achieve a higher price and often involves differentiation at the symbolic and the material level.[1] In the case of farmed salmon, however, profitability has relied almost entirely on economies of scale. And, because it has by and large succeeded in producing more salmon at a lower cost, the Norwegian salmon farming industry has been surprisingly profitable for decades, in spite of a steady decline of salmon prices on the global market (Asche, Roll, and Tveterås 2007).[2]

Aggregated statistics about Norwegian salmon aquaculture tell us that during four decades, from 1971 to 2011, the production output of farmed salmon increased by a factor of ten thousand (from nearly 100 metric tons in 1971 to just over 1 million metric tons in 2011; see figure 6).[3] And the growth has been steady: During the first four years in this period, the production grew tenfold (from 146 metric tons in 1972 to 1,431 metric tons in 1976; Statistics Norway 2012). During the next fourteen years, the production of salmon grew tenfold again, reaching 145,990 metric tons in 1990. This was followed by a brief period of crises and restructuring in the industry, after which production increased again. From 1992 to 2003 the production nearly quadrupled, then it temporarily stabilized at just over 400,000 metric tons until it peaked again at half a million tons in 2003. Since then, there has been a steady increase, until it peaked at more than a million metric tons in 2011.[4] Since then, the growth has continued. These numbers make Norway, with a population of only 5 million people, not only the leading producer of farmed salmon but the second-largest exporter of fish and fishery products after China (FAO 2008, 48).[5]

More than a million tons of salmon a year is equivalent to 12 million salmon meals a day. It involves 163 trucks packed with fresh salmon leaving Norway *every single day* and equals more than four times the total production of meat in Norway.[6] Scalable? Absolutely. A transformative turn in food production? Yes, indeed. Salmon farming has been the model for attempts to domesticate a variety of other marine species, such as halibut and cod, none of which have been nearly as profitable as salmon. But transformative potentials can be detected only in hindsight. So

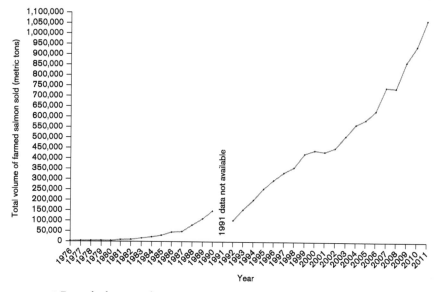

FIGURE 6. Farmed salmon production in Norway, total volume 1976–2011, generated from sales statistics (Statistics Norway, Table 07326: Akvakultur: Salg av slaktet matfisk, etter fiskeslag, https://www.ssb.no/tabell/07326/)

when the first Norwegian salmon were placed in a netted enclosure in the late 1960s and hand fed with fish scraps ground up in a cement mixer, the venture was hardly the result of ingenious foresight. Rather, as Anna Tsing (2012a) argues in the case of plantations, these salmon and their caretakers "stumbled into history and only afterward became the model for further scalable designs" (510).

Salmon aquaculture was never designed for this kind of growth. On the contrary, the idea in the beginning was that salmon aquaculture would be a useful source of secondary income for small dairy farmers on the steep slopes of Norwegian fjords, who had already been marginalized by more effective agricultural production in the eastern parts of Norway. Hence, the Norwegian aquaculture policy of the 1970s and 1980s placed an upper limit on growth and required that aquaculture enterprises be locally owned. As a result, aquaculture investors turned to Chile and Tasmania, while profit, by and large, remained local (see also chapter 2). In the early 1990s, these regulations were lifted, but their impact left a mark on the structure of the industry. Today the size of salmon aquaculture firms in Norway varies from large multinationals to relatively small family-owned operations. This historical trajectory reminds us that nonscalability can be written into a design of any enterprise and that capitalist expansion is far from inevitable (see Polanyi 2001), but also that it is a matter of politics.

So how can we describe the recent expansion of salmon farming without being trapped in a too-smooth account of industrial success? Such stories can be triumphant (such as Osland 1990), or they can be critical (such as Berge 2005), but they often overlook conditions that are beyond the scope of human intentionality and pay less attention to outcomes that are unpredictable, fragile, or unexpected.

I want to tell the story of expansion in a way that remains attentive to the specific forms and relations that emerge in ongoing practices of domestication among lively beings and things. Just as my daily tasks at Vidarøy quickly became routine, a set of repetitive embodied movements that required only minimal attention, so too do stories of salmon aquaculture get routinized, created with a limited set of narrative plots (see also chapter 1). The ethnographic challenge, then, is to move slowly, look more closely, and to ask these questions carefully and repeatedly: Who and what populates these relations of domestication, and what are they made of? And what are *their* contributions to this puzzle of expanding biomass?

Slowness (see Stengers 2011) is about methods, but it is also about positioning: it is about aligning oneself with the temporalities in the field and letting them guide one's focus. In this chapter, I draw attention to the temporalities of textured attachments, and let them set the pace.

TEXTURED ATTACHMENTS

The salmon domus is an assemblage of elements, all potentially transforming and shaping one another. Scalability can be conceived of as a kind of expansion, or the ability of a system to be upscaled or to accommodate growth. My concern here is the relations that hold the growing assemblage together, its inscribed temporalities of attachments and detachments, and how this relates to scalability.

In order to talk about the spatiotemporal choreographies holding this fragile, yet expansive regime together, we need to attend to their attachments. The term *attachment* alerts us to that which holds and that which is being held; it speaks of a relation of proximity or affinity, without a precise boundary and with no specified direction. Thus, it helps us see that through salmon farming practices, particular configurations of humans and nonhumans come together in a way that engenders translations, or transformations that make the assemblage durable and strong. Latour (2005) has argued that when it comes to building strong assemblages, "the more *attachments* . . . the better" (217, emphasis in the original), but, as I shall argue, detachments are just as important; the point is how the two work together.

I shall also draw attention to the textures of such attachments and their temporal choreography. Growth and expansion are spatial metaphors, and scalability is often conceived as spatial expansion (see, for example, Law 1986). But it can also involve a change of pace. This chapter is an attempt to think about scalability and growth as a matter of speed, suspension, and repetitive cycles, and to ask what difference it makes when temporalities are taken into account.

In what follows, I offer four ethnographic vignettes that highlight particular phases in the salmon life cycle when growth relies upon specific attachments and detachments. Let us start in the early stages of a salmon's life, as it unfolds in the dim, cool, quiet premises of a salmon hatchery.

Growing Salmon 1: Water, Eggs, and Degree-Days

John and I have driven many kilometers on narrow, icy roads to arrive at Idunvik, the hatchery and smolt production site where we occasionally spend a few days working with the people who work with the baby salmon. Cranes tower over two-story warehouse buildings. Large new cement tanks are half covered by wooden walkways, and the plastic pipes are being fitted by temporary workers wearing hard hats and logos we don't recognize on their overalls. For more than a year now, they have shared the space with salmon and their people. Soon this site will double its production of smolts and supply fry for other companies too. Salmon prices are good, and there is plenty of money for investment. Last year, 10 million fry came out of this facility. Outside, the construction noise is constant, and any conversation requires an effort. But inside the roe-incubation facilities, it is quiet, almost silent. In the chilly, humid, and dimly lit hall downstairs, the only sound we hear is the water flowing steadily through each incubation tray, holding eggs almost ready to hatch.

Rivers are breeding sites for Atlantic salmon. Spawning females dispose thousands of eggs in the gravel riverbed, some of which are immediately fertilized by males, who compete to position themselves next to her. Biology textbooks teach us that eggs spawned in the fall become fry the following spring. The incubation period—which transforms fertilized eggs to eyed ova and then to alevins—is primarily a function of water temperature and is expressed as degree-days (temperature × time). According to handbooks of salmon farming, fertilized salmon eggs hatch at about 475–500 degree-days (Stead and Laird 2002); in Norwegian hatcheries, this typically translates to around 60 days at a temperature of around 8°C. Once they hatch, they are called alevins in English (*yngel* in Norwegian) and are 10–15 millimeters long. Alevins feed from their yolk sac, which serves as an endogenous source of food for another 250 degree-days until they reach about 25 millimeters and emerge in the river as fry, able to feed themselves (Verspoor, Stradmeyer, and Nielsen 2007, 31). In hatcheries, the alevins remain in their Astro-Turf trays until they are transferred to tanks, at which point they are referred to as fry in English (in Norwegian they are still called *yngel*). This transfer is also the time for their "first feeding" (see the section "Growing Salmon 2").

A salmon hatchery mimics a river through the careful alignment of eggs and the flow of cool water. The journey from fertilized egg to fry is estimated to take about 800 degree-days. In hatcheries, this usually translates to about 100 days at around 8°C. Hence, the cool water is not only a "container" in which salmon can grow but an intrinsic element of growth itself.

But the eggs are not alone. Every day we check the temperature, making sure it is steady and cool. Every day we check the calendar and calculate, anticipating the day they are likely to hatch. And every day, we examine every single tray with roe, seeking the white ones and gently sucking them out with a transparent plastic tube and a siphon. Dressed in white rubber overalls, raincoats, and rubber gloves lined with cotton, equipped with a bucket and a siphon, we spend the days attending to roe, literally separating live matter from dead matter. Most dead eggs are visible as light-colored balls against the live orange ones. But some are in between. These I pick up and show to Tone, who has taught us how to do this, and she confirms: "*Misfostre*" she calls them—deformed fetuses—and even if they are alive, they need to go.[7]

Working with a siphon is new to me, and I feel clumsy. Closing my mouth around the plastic tube, I direct its lower end into the tray and aim for the grayish debris, the dead matter that needs to be removed. Sometimes water flows into my sleeve. Sometimes the vacuum I make is too strong, and I see perfectly healthy eggs come shooting up the transparent tube; then I quickly release my breath and start over. When I succeed in sucking up the dead ones, I immediately press my thumb firmly against the opening of the tube, holding them steady until I point the tube into the bucket and release. But my thumb doesn't seem to hold tightly enough, and time and again dead eggs slide back into the tray. After a while I discover that my tongue can close the tube much more effectively than my thumb, and as I breathe slowly in and out through my nose, I can hold the eggs in place however long I want. But occasionally I suck too hard, and feel a splash of cool water or unidentifiable debris on my tongue. I try not to swallow.

There is a silent nearness, short of words. Lips—closing tight; cold water on my tongue. Breath—holding it and letting it go in patterned cycles. Hands—steering the end of the plastic tube to the precise location of the imperfect egg. Eyes—tiring in the dim light. I often make mistakes. My limbs and senses are only loosely coordinated, and less so, it seems, than those of my coworkers. An egg splashes back in the tray and disturbs the quietness of the hatching ones underneath. A swarm of orange live matter floats up for a few seconds before the surface calms down again. How careful do we need to be? Who knows? As the operation manager, Olav, later remarked, we need to find a balance between cleaning the trays properly and not disturbing the eggs too much. My suggestion to put up some more light to make the job easier is rejected on the grounds that the vet wouldn't approve. "The eggs prefer the dark," Olav says.

Many things are going on here. There are alignments on multiple scales; there are textured attachments, but detachments too, enacting distinctions that are consequential: Tray or bucket. Life or death. When we are done, the trays look more uniform, homogeneous, pretty pearls of orange, at least for a little while. But their liveliness will continue to produce gray debris. And my clumsiness is as consequential as my skill. The genetic variation in the tray is perhaps just a touch different than it was in the morning. But exactly how our siphon-enabled encounters made a difference is hard to say.

The calculation of degree-days, so crucial in timing the development of young salmon, could appear as a scalability device. In principle, one could imagine doing everything more quickly, thus producing more fry per unit of time. However, as salmon farmers soon learned, this is hardly advisable. Experiments have shown that there is an ideal range with an upper threshold. If you speed the process too much, deformities will occur more often. Hence, the speed of incubation typically varies in the temperature range of 8–10°C (see also Stead and Laird 2002, 69).

Attending to the hatching roe, I am reminded that control is hardly realized and that "human mastery" is mostly in our heads. Fish are elusive; our relations are partial, fleeting, and contingent. In our trays, as at the bottom of rivers, liveliness presents itself as failure: shadowy hinterlands of not-quite-realized life-forms are endlessly produced.

A hatchery is, in principle, just a heterogeneous construction that aims as far as possible to mimic what happens in the river, through the careful alignment of fertilized eggs, rugged surfaces, and cool water flow.[8] Scalability is achieved by multiplying the number of trays or tanks (which requires more staff, more capital investment). The whole process may be sped up or slowed down through subtle variations in water temperatures (within a prescribed range). A slight increase in temperature will speed up the development of eyed ova, so that one may expect them to hatch not on a Sunday but on the preceding Tuesday. But for the eggs, there aren't any days yet, just the steady running water and the dim electric light in which they grow. In these dark rooms of silent anticipation, there is no outside or inside, no clear boundary between the salmon and the domus, just a series of temporal attachments and detachments that together constitute the very beginnings of a new salmon life cycle.

Narratives of domestication can be told as a historical journey that *separates* nature from culture. However, inside a hatchery, there are no clear boundaries between salmon and surroundings, no relata preceding its relations (see Barad 2003), no nature separate from culture. This is the nature of salmon domestication in its early stages: a site and a moment where specific *life-forms* (very young salmon) emerge together with particular *forms of life* (industrial labor practices) through carefully choreographed alignments of hands and siphon, siphon and lips, lips and water, water and eggs.[9]

Images of control, or of human mastery over nature, are not entirely wrong. Salmon hatchery workers *know* something about salmon (for instance, about degree-days), and this knowledge makes a difference. But images of human mastery fail to account for the question that I try to hold as a guiding puzzle for this chapter: How is scalability achieved? What are the relational attachments and temporalities through which salmon are currently being done as scalable?

I draw here on a long tradition in STS of engaging the way in which artifacts are fitted together in order to understand how projects become scalable and expand. John Law's (1986) analysis of the significance of sixteenth-century Portuguese sea vessels for European colonial expansion is a classic example. He notes how, for instance, their resistance to attacks, their relative independence of their surroundings, and their versatility in the face of different weather conditions all contributed to their mobility, durability, and capacity to exert force. These were qualities that—combined with a robust system of ocean navigation—were crucial for the Portuguese system of long-distance control. Specific forms of detachment and realignment in relation to features of the vessels' environment are crucial in the analysis, and the novelty of this approach was precisely the attention that was paid to such seemingly passive agents at a time when military success was more commonly attributed to human strategy and ingenuity.[10]

This approach resonates with Anna Tsing's more recent approach to scalability as a feature of design (2012a, 507). As she points out, scalability is not a feature of nature, but the outcome of heterogeneous, and sometimes contingent, sets of relations. She also reminds us that scalability is always incomplete and that nonscalability is usually in operation as well, but somehow suspended in day-to-day operations, or even marginalized in stories that highlight particular trajectories of growth. Scalability, then, is possible only if the elements involved "do not form transformative relationships that might change the project as elements are added" (Tsing 2012a, 507). Keeping separate is key. Hence, as I shall argue, scalability is also about detachment, and about the dynamic choreographies through which detachments and attachments unfold in spatiotemporal cycles.

Growing Salmon 2: First Feeding and the First Enactment of a Batch

So when does the cycle begin? When does a young salmon first appear as a figure that makes it manageable as biomass? I identify this moment as the time that is described locally as the "first feeding." If the life of a salmon is all about putting on weight, then the first feeding is obviously significant, as it marks the transformation from eggs to biomass. But it is significant in other ways too. We are back at Idunvik—but in a different building and at a different time:

> Yesterday, John and I spent the entire day emptying trays of AstroTurf filled with tiny, lively alevins into fifty cylindrical tanks located in a building that is referred to as kvithallen, or the "white hall."[11] The alevins have made a big move; first from the AstroTurf trays inside the hatchery building to temporary storage in green square

basins placed on top of each tank with a drain at the bottom, and then, after having been weighed and counted, into the tanks proper, in batches of about 30,000 fry in each. The transitions were not without risk. To remove the AstroTurf from the tiny bodies, shorter than my fingernail, was difficult. And to move them again from the drained basins to the tanks was difficult too. We had to empty the water first. My eyes are tired from trying to spot them in time to catch them with a dip net before they follow the circular merry-go-round of a current down the drain and disappear from sight into the filters in the basement.[12] . . .

Now we are looking down into swarms of fry moving at the bottom of the green tank like shadows—or like bees, says Olav. We stay and watch for a while. He is clearly fascinated, and so are we. I lift my hand, and a green opening emerges in the middle of the black dense swarm: their movements seem coordinated as they spread out for a moment and move away, just to cluster again near the bottom a few seconds later.

We tiptoe around the tanks with rubber boots, speak softly, and use a dimmer on the light switch.[13] The fry gather near the bottom of each tank, crowding together for protection—or that is what people say. It is easy to imagine their fear. The wide, fluid expanse of their new surroundings must be overwhelming. But being crowded together means that they might be deprived of oxygen. They need to become more courageous and start moving about.

About 0.2 grams each, these alevins can no longer feed on their yolk sacs. From now on they will need to eat commercial fish feed—at first in the form of fine dust. If everything goes well, they will soon begin putting on weight. But will they eat? The transition is a moment of anxious concern:

> I pick up the dustlike feed between my fingers and sprinkle it over the tank, like salt.
>
> "Careful, not too much!" It is my mentor, Tone, who keeps an eye on what I am doing.
>
> Nearly all the fish are still on the bottom, but as we feed, a few start to come up, inquisitive, exploring. Some are remarkably fast, shooting after one tiny parcel of dust after another, while the rest are drowsily squirming together on the bottom. In the beginning, I watch carefully to make sure that some of the feed disappears before I move on to the next tank. It can take several minutes before the fish get interested. But after a few tanks of moving rather slowly, I decide pragmatically on a number: "three pinches for each." I sprinkle the feed over the entire surface three times and briefly notice how the pattern of fry differs in shape from one tank to the next before I efficiently move on to the next tank.

But Tone does it differently:

> I realize that she works more slowly, her gaze patiently focused on the tank in front of her, and I notice that unlike me, she is really paying attention. Once I have come

FIGURE 7. Will they eat? Alevins ready for their first feeding (photo by John Law)

up next to her, as quietly as I can, she starts using a soft, high-pitched voice as if she is calling their attention. It makes me think of the voice that women sometimes use to speak to newborn babies that cannot yet grasp the meaning of words: a voice of care, concern, and wonder, responsive and ready to articulate any visible response from the little ones. In this case, the little ones may be invited to "come up" *(koma opp)*; they may be invited to "eat" *(eta)*; and these seemingly trivial fish movements become articulated and meaningful as if they fill her with excitement and awe. She shares these moments with me whenever I show interest, and now and then we stand together for a bit, more intimate than we really are, both in strangely familiar terrain.

I have known Tone only a few days on this first visit to Idunvik, but it feels as if I have known her a lot longer. Facilitated by the fact that we are roughly the same age and by the unlikely coincidence that she went to school with one of my best friends, we have already shared quite a few cups of tea and waffles and exchanged glimpses of our rather different lives. Her work life has revolved around the hatchery; she has seen it grow and change ownership more than once. Concluding the fieldnotes for the evening, I write:

> This is not a maternity ward, and we are not best of friends. What I notice is simply this: that in this practice, there is a kind of human-to-animal "bonding" going on, or perhaps rather a relation-in-the-making, however temporary, in which there are elements of care and affect on the human side, and who knows what is going on down below. And then that the practice is gendered.

Voices, bodies, movements, affect, and quiet awe. Just as in the cases of hand-feeding out at Vidarøy, we see that practices of care are heterogeneous, embodied, and sometimes verbalized too. Enactments of care? There is no doubt about it. Affect? Plenty of that too. Again, we can recognize a relation of what I have referred to earlier as partial affinity, a creative mode of attunement that is embodied but never complete.[14] Like the male farmworkers' "checking the feeding," Tone responds to the fish by drawing on a collage of practices. In her case it includes practices of care, some that are fish-related and some that are not. Perhaps her care becomes effective because she allows her imagination to drift affectively between tiny babies and tiny fish.

But from the perspective of the firm, it doesn't really matter why Tone and her fellow workers in the white hall are attentive. The point is that they are, and their practices facilitate or guarantee the precarious transformation from alevins feeding from their yolk sacs to tiny salmon that can be fed by us. This moment of fragile transition marks the beginning of a journey, which is all about putting on weight. Salmon appetite is crucial, not only as a mediator that enacts batches of salmon as doing well or not-so-well through practices such as checking the feeding, but also because feeding, eating, and being fed are the intertwined practices of human and salmon that in daily interaction, as well as financially, sustain the salmon domus over time (see chapter 3). The transition to these tanks also marks the first time that the salmon are inscribed as numbers. Consider these fieldnotes, from the day before the first feeding:

> We don't know how big they are. And we don't know how many they are. All we know is that in a couple of days, they will be distributed into fifty tanks (about 170 cm across, 130 cm deep), along four different rows in the white hall. Each tank is supposed to have about 6 kg of salmon biomass. What follows is not only a complicated and careful procedure of moving tiny fragile alevins from A to B but also a series of calculations. The first thing we need to know is average weight. Linn begins by calibrating a bucketful of water on the weight so that the scale shows zero. Then with a dip net, she picks up a random number of fry, and slowly, while she counts each one, she releases a few at the time into the bucket. Once she has reached the number 100, she weighs the bucket again. The difference appears as a digital number in the display: 21 grams. Divided by 100, we have now decided that each salmon weighs on average 0.21 grams. The procedure is repeated for salmon from three different provisional tanks, and in the end we decide 0.2 gm is a reasonable average weight for the alevins in the entire batch.

It is a trivial sampling, but it is the first step towards producing biomass, the first approximation of the numbers that will later be calculated continuously to produce the Feed Conversion Rate (FCR). But knowing their weight is not enough. We also need to know approximately how many we have placed in each tank. More calculations are needed:

> If each tank should contain approximately 6 kilos of fish biomass, and if they are just over 0.2 grams each, then a little less than 30,000 fry should go into each tank (6000 g divided by 0.21 g is equal to approximately 30,000). To prepare the physical move, we start exploring the bucket. How many kilos of fry will a full bucket hold? After a bit of fiddling back and forth, we decide that one bucket full of fry weighs about 3 kilos. That means it contains around 15,000 fry. Thus, two full buckets go into each tank. We line up buckets. Place them on the weight. Calibrate the scales at zero. And then, gently with a dip net, we lift up the wiggling mass of fry, let the water drip off and let them out into the bucket until the scale shows 3 kilos. And then we empty the bucket in the tank and repeat, twice for each tank.

Everyone knows that these values are approximate. Later, when they are distributed to bigger tanks, and perhaps also to other smolt production units, the salmon will be sorted automatically. Shortly after that there is vaccination, which provides a chance to count them again. But for now, an approximation is all we have and all we need. The reason we need it is to know how much to feed. More fieldnotes, a few days later:

> Tone is doing arithmetic again. Guided by a printed form posted on the wall inside the white hall with figures for expected growth for different sizes of salmon (she says it is a little old but still valid), she does the calculations by hand. According to the form, fish of this size are supposed to have a standard growth rate of 3.1 percent per day.[15] If each tank contains 6 kg of fish, it means that they need 186 g of feed each day. We are assuming now that their appetite is average, which is something we cannot judge yet.
>
> Tone: "It is so tempting, to start to give them more, immediately. But you cannot do that; you need to be careful. To feed too much will just clog up the system and not do them any good."

Until now, the development of the egg and the alevin has unfolded in the silent coolness of hatchery trays, as a relation between salmon and their fluid environment, mediated by temperatures (cf. degree-days). From now on, their development relies on a carefully recorded distribution of feed. We know they are not exactly 6 kg and 30,000 in each tank, but this approximate assumption is sufficient; it enacts them as a batch that is quantifiable and makes further growth possible. They have entered, and for the first time been "digested," into the calculations of the firm; they have become an entity that is legible, economically as well as biologically, and thus been rendered mobile. It is this transition that is the basis for further growth.

So far, we have approached the salmon as batches that grow. Each batch will eventually be translated to profit in a way that enables further investment and further expansion (see also chapter 4). But how are such expansions done? What are their material components? How do they work, in practice? Let us fast forward some 500 degree-days in order to realign with the young salmon as they have added about 5 cm to their body and been relocated to a much bigger tank in a different smolt production site further north.

Growing Salmon 3: Smoltification and Black Roofs

We are back in Frøystad, the smolt production site, which is more than thirty years old and referred to locally simply as *smoltanlegget* (see chapter 4). The operator, Tore, recalls the early years, when fish feed was mixed in cement mixers and made from fish scraps from local fishermen. The production site is near a small river, which has been fitted with a simple dam construction. In spring, the hissing sound of running water makes a steady noise around the outside tanks and walkways. The river used to be a salmon river, and some thirty years ago, rivers like this supplied broodstock for experiments with salmon aquaculture. Now, the salmon seems to be gone, but the river supplies freshwater for the tanks that hold salmon until they smoltify and can be transferred to the seawater pens in the middle of the fjord.

Salmon that begin their life cycle in a river remain in their river of origin as long as it takes to put on weight and reach the size and stage of maturity that is compatible with life at sea. The length of the freshwater phase varies not only with latitude but also within a population. In North Atlantic rivers, salmon typically spend between one and four years in the freshwater phase before they undergo the physiological and morphological changes associated with smoltification (Verspoor, Stradmeyer, and Nielsen 2007; Stead and Laird 2002, 80).[16] Smoltification typically occurs in spring, when the juveniles are around 10 cm long, and is triggered by the shifting light of the different seasons.[17]

But the global demand for farmed salmon pays little attention to such seasonal shifts, nor does the economy of the aquaculture industry, which relies on permanent employment of local workers.[18] In order to supply salmon for slaughter all year round, the production regime of the Norwegian aquaculture industry requires a steady flow of smolts. In practice, this translates to a pattern of smolt delivery twice a year, with each delivery stretched over a period of a few weeks: *vårsmolt* in spring, and *høstsmolt* in the autumn. So how do you make salmon smoltify in early autumn, rather than in spring?

Part of the answer is the erection of a cone-shaped black roof, which makes it look as if a house has been erected around the tanks. The seasonal change in length of daylight (photoperiod) from winter to summer is an effective trigger of smoltification of salmon, both inside the tanks and in nearby rivers. The black roof

FIGURE 8. The black roof at the smolt production site (photo by John Law)

effectively blocks this relation between seasonal change and salmon maturation cycles. As it eliminates the seasonally shifting daylight, it detaches the smoltification process from the annual cycle aligning the sun and the earth. Detached from the stimulus of shifting daylight and attached instead to electrical lights inside the dome, the salmon can be placed on a different trajectory so they will smoltify in the fall, rather than the following spring. Hence, they can be delivered nearly half a year earlier than their cousins in the outdoor tanks, and batches of smolts can be delivered nearly all year round. With a steady growth in salmon production over the last decades, this is now standard practice in commercial smolt production.

In practical terms, as the vet later explains, they separate what they call the one-year-olds (*ettåring*) from the zero-year-olds (*nullåring*). The one-year-olds, which have something similar to a natural cycle, are called *vårsmolt*, spring smolts. They hatch in winter, are transferred to the smolt production site in spring or summer, and are vaccinated the following autumn. Then they grow some more the following winter, until they smoltify in spring and are ready for sea around May. This production cycle can be done outdoors, as the longer days in spring will trigger smoltification.

The zero-year-olds go through what the vet calls a "steered regime." After hatching early in winter, they are taken through an "artificial summer" with full light twenty-four hours a day for about six months. In July, they are vaccinated, and then they are given twelve hours of light and twelve hours of darkness for about six weeks, while the Nordic summer night is light outside their dome (this is when the black roof is really needed). Then, when they begin to smoltify, the lights are again turned on day and night, until the smolts get ready for delivery to the grow-out site in the early autumn; hence, they are referred to as autumn smolts *(høstsmolt)*. Unlike the smolts delivered in spring, these autumn smolts spend their first six months at the grow-out site in winter, then grow through summer and a second winter before they are ready for slaughter in spring of the following year. (The spring smolts also spend a year and a half at sea, but they get two summers and only one winter.)[19]

In this way, physiological processes that are otherwise suspended through the dark winter months are triggered through the use of electric light. The black roof choreographs a temporal detachment, within which other forms of attachment or alignment can take place. As a feature of design, it enacts scalability through the simultaneous movement of detachment and realignment: *detached* from the seasonal changes at 60° north, salmon are instead *aligned* with consumer demand, or rather the articulation of such demand, through the network of slaughteries, freight carriers, spot-market prices, export arrangements, and emerging sushi restaurants around the world. The result is a doubling of the output of smolts from more or less the same available tank facilities. But the benefit for the salmon industry is also related to steady supply. In the words of Stead and Laird (2002): "By using out-of-season smolts, it is possible to spread freshwater production across the year, allowing greater annual output of smolts and better utilization of plant equipment and human resources. . . . Hence, the peaks and troughs in production can be reduced" (89).

The "steered regime" involves a deliberate interference with the photo environment of the young salmon. When it works, it can be seen as a specific sequence of attachments and detachments: an anticipated steady demand of a global market is literally built into the design of the assemblage through a detachment (the black roof) of the tanks from the light Nordic summer nights (or, we might say, from the earth's orbit around the sun). This detachment allows a temporal realignment of the operation, from the temporal cycles of the Nordic seasons to a chain of production in which market demand is imagined as noncyclical.

Each tank and each new batch of smolts have come to resemble what Anna Tsing playfully refers to as a "pixel": that which remains uniform while the image, project, or assemblage is made bigger or smaller. Just as the scalability of an image depends on the *autonomy* of each pixel, so too does the scalability of the salmon farming operation depend on the relative autonomy of each tank. The detachment of the tank from local seasons ensures, or produces, a specific version of scalability. But while pixels do their work in relation to space, the scalability here works in relation to time, so that a batch of smolts becomes like a temporal pixel.

While smoltification is triggered by light, subsequent growth is facilitated through feeding, which brings us to another way in which scalability is done. Let us fast-forward another year, to the grow-out site, where the salmon's job description is simply to put on weight.

Growing Salmon 4: Peruvian Anchovies and the Extraction of Water

Feed and salmon. Salmon and feed. This is one of the crucial relations holding salmon assemblages together, and one that—as I shall argue—has enabled aquaculture enterprises to expand far beyond what anyone had predicted. Salmon are enacted as hungry. They eat, they gain weight, and weight translates to price and profit. Putting on weight is their job description, and feed is an important part of the enterprise. But that does not mean that the firm, or indeed the caretaker, can speed it up.

Recall Kristoffer as he climbed up on the ramp to watch the salmon eat (*sjekke foringa*) in chapter 3 ("The Water Surface as Interface"). While the feed distributor was running, he threw pellets onto the water surface to check their appetite. Sometimes they ate a lot. Sometimes they hardly ate anything. If they're not eating, it's worrying, but there may not be a whole lot you can do about it. After all, salmon aren't like geese. You're not making foie gras: you cannot force-feed a salmon. So, for Kristoffer and others who work on the fish farm, feeding salmon is a process of constant adjustment, of watching attentively, of responding to what they sense or see and trying something different if that doesn't work (Mol, Moser, and Pols 2010). As John Law and I have detailed elsewhere (Law and Lien 2013, 2014), it is about tinkering; it is about care.[20] But it is also about securing access to marine resources.

Salmon are carnivorous—they feed on other fish. But some claim that other fish are getting scarce. This means that in ecological terms the growth of salmon farming could be unsustainable. In economic terms, feed is getting costly. In geopolitical terms, Norway needs to secure access to feed resources for its industry to thrive. The whole assemblage is intricate and fragile. It is a balancing act through which economic, environmental, and political concerns, as well as fish appetite and human nutrition, are constantly being played off against one another.

Salmon are bred to be hungry. In a single week an adult salmon will eat about a quarter of its own weight. Aquaculture stocks are evaluated in relation to their companies' reported feed conversion rates (see chapter 4). Feed must be efficient. Feed supply must be steady and predictable. This means that feed must sometimes travel great distances, and it also needs to be stored. A site with more than half a million fully grown salmon will need nearly 50 metric tons of dry pellets—every single week. That is a lot of pellets. So how is this done?

At the floating structure called Vidarøy, the largest room is the warehouse that serves as a feed storage room. With its high ceiling and convenient location near the silos, it is perfect for storing feed. Every three or four weeks, it is filled with feed bags unloaded from the feed boats that serve one of the three major feed companies (EWOS, Biomar, and Nutreco) that crisscross Norwegian fjords every day, to and from localities where salmon grow. The room can hold about 200 metric tons, and after each feed boat delivery, someone will need to spend the entire day on the forklift, stacking the bags up to the ceiling.

What are pellets made of? Most people who feed salmon don't know. Their focus is on how much they eat and what their feeding behavior reveals about their state of health. Very few are interested in what goes into the pellets. Environmental activists, on the other hand, are very interested, and some consumers are too. Most likely as a response to such interest, one of the main suppliers, Skretting, has produced a video and published it on its website. The video promotes its traceability system and gives a fairly detailed breakdown of where the fishmeal and the fish oil come from. In practice, the answer varies with shifting availability, but in Skretting's example, which is a batch of actual pellets shipped out from Stavanger, Norway, about half of the fishmeal is based on Peruvian anchovies. The other half consists of different fish and trimmings, caught in the North Sea (see table 2).

To feed millions of Norwegian salmon with fresh Peruvian anchovetas and North Atlantic Ammodytidae, or *tobis* ("sand lances," or "sand eels") would be an unimaginable task. Multiply the anchovetas, and imagine piling them on top of each other—they would be a messy heap of rotten fish. Their nutritional value as fish feed would disintegrate. They would decay, facilitated by millions of bacteria that thrive in the mushy, humid, smelly pile of rotting fish. However, bacteria need a minimal amount of water. By transforming wet anchovies into dry pellets, the processes of decay are suspended. Hence, integral to the salmon feed-farming assemblage is another process of detachment: the extraction of water from the fish substance. Extraction of water ensures the suspension of bacterial processes of decay.

I suggest that it is precisely these detachments that make salmon aquaculture scalable. Folded into these tiny dry feed pellets is a giant "time machine" that enables the massive movement of feed between the South Pacific and a Norwegian salmon farm. These detachments (of water from anchoveta and tobis, or sand eel substance) make durable and mobile the key ingredient that is needed for salmon to grow in the domus (cf. the Portuguese vessels in Law 1986). Without this mechanism for "freezing time," or for suspending decay, the globally expanding salmon farming project could hardly have been possible. Similar pellets feed chicken and pigs, by the way, with similarly scalable effects.

TABLE 2. Origins of fishmeal in feed pellets, kilograms

	Chile	Denmark	Iceland	Norway	Peru
Anchovy	818,449	—	—	—	6,384,451,007
Blue whiting	—	29,163	154,419,392	2,747,548	—
Boar fish	—	417,316,921	—	—	—
Capelin	—	—	299,007,655	3,417,411,240	—
Capelin trimmings	—	—	50,615,876	7,378,896	—
Herring	—	—	138,702,990	11,453,077	—
Herring cuttings	—	4,265,271	2,275,458	510,069,749	—
Horse mackerel	818,449	—	—	—	—
Mackerel	—	—	76,817,068	—	—
Norway pout	—	186,973,404	—	37,509,243	—
Sand eel	—	119,122,513	—	49,376,551	—
Sprat	—	3,613,490	—	15,092,924	—
Other	409,225	—	—	1,526,496	—
Trimmings	—	—	619,704	273,418,615	—
Total	2,046,123	731,320,762	722,458,144	4,325,984,337	6,384,451,007

SOURCE: Skretting, "Marine Resources" (eTrace table at 2 min., 30 sec.), *Tracing the Truth*, YouTube video, uploaded January 13, 2011, http://youtu.be/J3ONqWXYe18.

NOTE: The composition of feed is subject to fluctuations.

We could refer to the feed pellet as an "immutable mobile" (Latour 1987). What is being done here is, in effect, a third form of mimesis of processes "in nature." But rather than mimicking salmon lives in the river, as in the case of hatching eggs and smoltification, the feed-pellet mimesis is more generic. Detachment of water from dry matter is similar, for example, to what happens in deciduous trees in the fall, when water retreats back to the roots and the whole organism comes to a standstill. What the feed companies have done is to take a familiar biological process of seasonal and cyclical fluctuations of water and tweak it for other purposes. In this way, domestication involves not only the fish itself but an entire feed-fish cycle. Unlike carp, which have been kept in ponds for millennia and feed primarily on plant material, salmon are carnivorous—at least for now. Experiments to feed salmon a more vegetarian diet have been fairly successful, but plant materials are only partially replacing the marine feed. Hence, salmon-farming industries remain fundamentally dependent on marine feed resources and thus deeply aligned with global ocean resource extraction. In this way, hungry salmon in the North Atlantic (as well as those in the South Pacific, such as Chile) are mediators in a complex chain of connections that link the consumption of smoked salmon in Asia and North America to the daily lives and subsistence of fishermen in the South Pacific. The networks are extensive, politically significant, but also slightly beyond of the grasp of this ethnography. What I wish to highlight here is the role of the feed pellet in mediating these relations, and specifically the role of the pellet in making salmon farming scalable.

The pellet is crucial for aquaculture expansion, but it is not enough. Trade agreements (for example, between Peru and Norway) are also needed to maintain this assemblage, and the composition of nutrients must be fine-tuned too. But none of this would work without the basic features of the pellet. The water-extracting technology described above is essential to the scalability of contemporary salmon aquaculture; without it, aquaculture simply could not expand. By suspending the processes of decay, massive amounts of marine resources are rendered mobile, and the massive and regular transfer of feed resources from the South Pacific to the North Atlantic is facilitated in a manner that resonates previous moments of colonization, such as the Portuguese expansion (Law 1986). This mobility sustains and stabilizes a new set of commodity relations, linking Norwegian salmon farming to salmon commodity markets from continental Europe to the Far East, while at the same time disembedding Peruvian anchovetas from their more immediate sites of fishery production. Hence, intrinsic to the scalability of salmon aquaculture is a very basic feature of design, the extraction of water, which in effect forges specific detachments that are essential to expansion.

The networks that are mobilized to grow salmon extend halfway around the globe. Domesticating salmon is a multispecies relationship, one that involves not only salmon but other fish too. But the effects are deeper, because enrolling species such as the Peruvian anchoveta means enrolling entire marine seascapes, each of which has its own unique human and nonhuman entanglements. As with other species, the domestication of salmon involves the domestication of seascapes and landscapes as well, some of which are geographically distant from where the salmon's lives unfold (see also Swanson 2013). Hence, the implications of domestication do not always meet the eye.

ALASKAN EXCESS

When sockeye salmon reach the shores of Bristol Bay, Alaska, they come in huge numbers. Karen Hébert (2010) tells how, for a few busy summer weeks, the fishermen haul in as much as they can and process it as quickly as they can before it goes bad. The rest of the year is less busy. Fishermen in Bristol Bay struggle to reinvent themselves in order to capitalize on the "wildness" of their product, which tends to lose out in competition with the cheaper farmed Atlantic salmon from Chile.

Alaskan salmon was once the backbone of a successful export industry. Since the early twentieth century, the shores of Bristol Bay have attracted fishermen, laborers, processing companies, and distant investors, whose fraught and fragile relations made Alaskan salmon into a profitable global commodity. Like farmed Atlantic salmon, Pacific salmon was scalable too. Hatcheries along important salmon rivers secured large returns of pink and chum, and it is estimated that about 31 percent of the salmon harvested in Alaska comes from hatchery production (Grant 2012, 325).[21] In the busy weeks of summer, the salmon returning to spawn along the Alaskan coastline are indeed abundant. The difference between the farmed salmon of western Norway and the Pacific salmon of the Alaskan shore is often portrayed as one between domesticated and wild, or between farmed and caught. But rather than reproducing this binary of nature and culture, we need to consider the various attachments and alignments that made each product scalable. When we do that, unexpected similarities appear, but also new differences.

An intrinsic and essential feature of the scalability of current industrial salmon-farming practices is, as I have argued, the feed pellet and the extraction of water, which suspends processes of decay. These processes allow "Peruvian anchovies" to travel halfway around the world and then to be distributed and stored in Norwegian warehouses without forming "transformative relationships" that might change the very structure that makes their nutrients palatable and available for salmon's digestive biology.[22] Farmed Atlantic salmon stay put, but the feed they need is brought to them from oceans far away, partly as a result of this water-extraction technology.

Alaskan salmon, on the contrary, do *not* stay put and they are *not* fed pellets to gain weight. Instead, they source their food in the Pacific Ocean, thousands of miles from the river where they spawned. The species they feed on may be similar, but unlike farmed salmon, they transform the fish they feed on "on site," so to speak, into storable fat and proteins, which in turn become the "wild salmon" flesh that humans find attractive. Hence, the salmon that return to the shores to hatch have effectively fed *themselves,* and the problem of storage and transport of feed is thus conveniently solved. All the fishermen have to do is to scoop them up as they arrive.

However, as Karen Hébert (2010) shows, this is easier said than done, because like "Peruvian anchovies," salmon flesh is perishable. Hence, salmon require careful handling and processing in order to suspend the processes of decay that could destroy the structure by which salmon flesh becomes palatable and available to the human digestive system. Because the reproductive cycles of Alaskan salmon are finely attuned to seasonal cycles, the salmon return in summer more or less all at once, making the logistics of processing and handling all the more difficult. How do you handle such huge amounts of salmon in a way that renders it palatable for distant consumers and keep it stable at the same time? The solution? It is, quite simply, the sealed can.

The technology of the metal can, like the feed pellets, can be seen as a "time machine"; this one enabled Pacific exporters to suspend the process of decomposition and thus to make salmon available to consumers all year round. The can made salmon storable. Today as Alaskan fishermen struggle to reinvent themselves to become competitive in a global market where freshness is becoming a key value, it appears somewhat old-fashioned (Hébert 2010). But for almost a century, it served its purpose well. It achieved what the food pellet achieved in the context of salmon farming: it "froze time." Through the combination of pasteurization and metal sealing, both of which render the contents of a can sterile, processes of decay are suspended. We could say that the ready-to-eat salmon is detached from the potential for harmful bacteria that would otherwise interfere. The effect is similar to that of the feed pellet: a product that can travel and that can also be stored. No longer at the mercy of the seasonal cycles of salmon migration or the inevitable cycles of biological decomposition, Alaskan canned salmon and Norwegian farmed salmon are industrial products of different eras but with similar end results: global mass commodities. One is caught, the other is raised, but their similarities, as well as their differences, have less to do with "wildness" than with the specific materials through which temporalities of decay are suspended. Farmed salmon rely on pellets. Alaskan wild salmon relied on a can.

SCALABILITY, TEMPORALITY, NATURE, AND THAT WHICH IS LEFT IN SHADOWS

Scalability, then, is not some abstract feature or the disembedded design of industrial capitalism or the so-called modern. Nor is it a property of a "thing" itself, independent of its material surroundings. Rather, it can be understood as a property of the relations, as well as the attachments, through which the assemblage is enacted—attachments through which specific temporalities are already inscribed, and through which new temporalities may be enacted, as in the case of autumn smolts. These attachments can materialize as different things, such as a tank, a light, a black roof, steadily running water, a feed hopper, a siphon, a pellet, and a Peruvian anchoveta.

The pellets are noisy as they shoot through the pipes above our heads, but the Peruvian anchovetas are silenced in the process, potential connections severed as they transform from wet fresh catch to dry stored pellets. With the water extraction, and the changing of hands through transnational transactions, origins are blurred; in the end, a fish-feed pellet is just that: a pellet, named for its purpose, rather than its past.

. . .

Temporalities reside everywhere. In the day-to-day practices of salmon farming, they multiply beyond the relations I have discussed: there is the salmon life cycle; there are the feed deliveries and the rhythm of the feeding; there are the hatching of eggs, the timing of slaughter, and the four-hour window between death and rigor mortis during which salmon can be handled after slaughter; there are the fluctuations of the market; there are the shifts of the farmworkers, the timing of salmon sexual maturity, and the annual and diurnal rhythms of humans and salmon. All this tells us that in practice, in every movement and every attachment, temporalities are already inscribed. It also tells us that in every arrangement different temporalities are being negotiated, tinkered with, and adjusted. Sometimes this tinkering fails: a salmon fails to grow to a prescribed size within a given time and is rejected. But even the rejects are the product of a larger choreography, which sorts fish according to size (Law and Lien 2013, see also chapter 4). And in this way, the farmed salmon are aligned again, this time with each other; they are scalable by virtue of being managed as a batch—calculable, predictable, and standardized.

In this chapter, I have attended to practices, attachments, and their textured temporalities; I have told the story of salmon domestication and industrial expansion in a way that is generative of a world in which nature and culture do not constitute a foundational opposition. Instead, I have described some selected spatiotemporal choreographies that enable or engender quite specific forms of attachment. I have tried to show that scalability is not only a matter of spatial expansion, but that temporalities matter too. It is when these are aligned in very specific ways that the attachments (and hence the assemblage) come to appear as durable, scalable, and strong.

Approached as a crucial transformation in the historical journey "from nature to culture," the domus is often conceived as that which locks nature out. And, indeed, the enactments of the boundary between the wild and the domestic are many: nets, walls, filters, calculations, maps, scale samples, and so on (Lien and Law 2011, Law and Lien 2014). The black roof covering the tanks could seem at first to be yet another such enactment of a boundary. But then, as we look closer, we find that inside this dome and in the dimly lit trays holding the fertilized eggs, another nature is brought into being, through the skillful switching of light on and off in particular patterned cycles or the gentle shifts in water temperature. In this world of finely tuned relationalities among salmon-light-water and growth, salmon come out as smolts, or eggs hatch sooner—or later—than they would have if they had been immersed in river gravel or feeding in rivers. But then again, it all depends on perspective. Because, when viewed from inside the dome, the autumn smolts emerge as silvery and ready for sea just in time, perfectly aligned with a temporal cycle that is no longer attuned to the earth's orbit around the sun.

To think of domestication as shutting nature out, or as enacting a boundary to the wild, is to capture only one way in which domestication is done. Another approach is to see seasonal shifts of daylight or the shifting temperatures or the extraction of water as features of design that, folded into the salmon assemblage, make the salmon domus a world in and of itself, and farmed salmon neither nature nor culture but both—or perhaps something altogether different.

6

BECOMING SENTIENT

Choreographies of Caring and Killing

How do we come to terms with inflicting death on animals? How do we deal with their suffering? The question might be as old as human beings. Recently it has been addressed from many different angles, from animal rights activism to animal welfare regulations. Most debates concern terrestrial animals that can walk up to us and look us in the eye. But as aquaculture industry expands, such questions are now being asked about fish (Huntingford et al. 2006; Damsgaard 2005; Lund et al. 2007).

Following the massive expansion of salmon aquaculture, European animal welfare legislation is no longer restricted to four- and two-legged, furry, and feathered companions. Farmed fish have emerged as sentient beings as well, and thus they are subject to welfare regulation on the assumption that fish such as salmon *might* have the ability to feel pain. Although the question of whether fish feel pain is still contested among biologists, it is sufficiently settled to give farmed salmon in Scotland and in Norway the benefit of the doubt.

But animal sentience concerns so much more than welfare regulations and animal neurology. It is also about care, compassion, or neglect in the day-to-day interactions between salmon and their human companions. It relies on practices that involve sophisticated monitoring technologies as well as a trained eye. All of this alerts us to the entire assemblage that I refer to here as the salmon domus, that heterogeneous gathering of human and nonhuman entities that defines, or enacts, what a farmed salmon is or what it can be. Thus, if care is enacted, that enactment is often already an essential element in the salmon domus assemblage (see chapter 3). Exploring sentience, we thus need to attend to the salmon assemblage as an enabling structure and ask what scope it allows for salmon well-being and what scope it allows for humans to care (Mol, Moser, and Pols 2010), or to tap into that emotional register of compassion that makes them responsible as well as response-able (Haraway 2008, 89). In other words, what is "care" in the practices of the salmon domus? And what scope does the salmon domus allow for salmon sentience to matter and for humans to care?

In this chapter I argue that farmed salmon in Europe are about to become sentient beings. For most practical, ethical, and regulatory purposes related to welfare, farmed fish are no longer "just fish": they have also become "animals," subject to animal welfare regulations. Recent legal moves have placed salmon sentience on the industrial fish farmers' agenda. They have led to the rebuilding of fish slaughterhouses and to mandatory courses in fish welfare in Norway. They have triggered, and also drawn upon, new research in biology on fish behavior and cognition, which asks questions such as these: Do fish recognize each other? Do they engage in social learning? Do they strategize? Do they cooperate? A yes to any of these questions indicates some kind of cognition and brings fish one step closer to being included in what veterinarians and animal rights proponents, following Peter Singer (1981), refer to as the "moral circle"—that is, "a collective of beings whose interests are given serious moral consideration for its own sake" (Lund et al. 2007, citing Singer 1981).

However, to approach sentience as a quality that nonhuman beings such as salmon either do or do not possess is a way of ordering the world that overlooks all the different ways in which sentience may be understood or enacted. While both philosophical and biologically grounded arguments in support of salmon sentience undoubtedly have made a difference in specific legal situations, they often fail to consider the heterogeneous relations through which farmed salmon come into being. I am thinking here specifically of the lively practices of growing salmon in tanks and pens. As we have seen in chapters 3 and 4, these relations are always more than just human and more than just animal too. Most importantly, they involve practices *in which sentience may be enacted* and in which human sensibilities (as well as philosophical texts and legal documents) exist as potential contexts, or framings, that may be evoked to a greater or lesser degree. Hence, I approach sentience not so much as a property of a salmon in and of itself but rather as a relational quality, a potential aspect of such heterogeneous relational practices. I suggest that it is in this space of uncertainty that the potential for improvement lies. In this chapter I will consider how relational entanglements hurt humans as well as salmon, and how pain can be lessened for both.

The rest of this chapter is a journey through different practices in which sentience unfolds. I begin with a brief review of how sentience is performed in the literature on animal rights and animal welfare and in biological research. I then turn to ethnography, exploring salmon sentience as it is being performed in salmon aquaculture practices. This section aims to show how sentience resides also in embodied practices, how it is relational and always specific. Towards the end of the chapter I give an account of sentience as it is legislated in Norway and in the European Union (EU) before I turn to one of the outcomes of recent legislation: mandatory courses in fish welfare. I conclude by considering salmon sentience in light of processes of domestication.

The scholarly literature on animal sentience extends from philosophy, sociology, and anthropology to biology and veterinary research. So what *is* animal sentience according to the literature? And where is it located? To provide some background for my subsequent account of the legal changes in European animal welfare, I will briefly touch upon some key topics in the debate. As will soon become clear, animal sentience is enacted, or argued, in widely different textual contexts, each with its own agenda. Hence, sentience can be many things, both literally and figuratively. Let us start in the realm of philosophy.

LOCATING SENTIENCE IN PHILOSOPHY: SUFFERING, SENTIENCE, AND ANIMAL RIGHTS

Since Jeremy Bentham identified the ability to suffer as the key criterion for granting rights to both human and nonhuman animals, philosophy has been an important source of inspiration and validation for animal rights movements. Bentham's famous statement from his 1789 treatise, "The question is not, Can they reason? nor, Can they talk? but, Can they suffer?" has been hugely influential in Western philosophical thinking about animal rights.[1] A similar perspective is advocated by utilitarian philosopher Peter Singer (1981). His assertion that the quest to seek "the greatest happiness for the greatest number" should include nonhuman animals has laid the foundation for animal rights activism as well as for more recent animal welfare legislation. Philosopher Tom Regan agrees that certain nonhuman animals have inherent value, but he comes to this conclusion via Kant's notion of the intrinsic worth or dignity of human beings, which implies that they should be treated as an end and never only as a means (cited in Huntingford et al. 2006, 339). Extending this notion of dignity to animals, Regan (1983) argues that since we are all "subjects of a life,"[2] we cannot justify disrespect for other subjects, whether human or animal; hence it is hard to defend, for instance, industrial farming for food. Other philosophers, such as Mary Midgley (1983), advocate an "ethics of care," while Bernard Rollin (1995) points to the distinctive species-specific nature of nonhuman animals as the basis for moral respect.

All these approaches can be seen as philosophically and culturally situated attempts to come to terms with some of the dilemmas that contemporary food production practices such as industrial farming pose for Euro-American orderings of society and nature, humans and animals. These are dynamic and perhaps less human-centered than they used to be, yet the dilemmas remain. As Buller and

Morris (2003) note, "While post-modernity has encouraged us to see the individuality and subjectivity of nonhumans as beings, modernity continues to put them on our plates as meat" (217). That we should both include animals in the "moral circle" and eat them remains a conundrum. This enigma lies at the core of the division between a mainstream plea for animal welfare and a more radical plea for animal rights, which for many activists leads to vegetarianism as the only possible solution.

Although none of these philosophers explicitly considers the status of fish when they speak of animals (hence it is often unclear whether their notion of animals includes fish), their arguments are engaged in recent advocacy for including fish in considerations of animal welfare (Lund et al. 2007, 3; Turner 2006). By comparing fish to other animals, and even more, by classifying fish *as* animals, a well-established literature on animal rights is enacted as a context within which a plea for fish welfare becomes meaningful. As we shall see, this "worlding exercise" (Tsing 2010) has been of great importance in the process of enacting salmon as sentient beings in the realm of animal welfare legislation.[3]

Enacting sentience in the realm of philosophy involves a reliance on forms of textual reasoning, where dilemmas are sought and solved by juxtaposing texts and arguments. A more embodied version of reasoning is offered by Donna Haraway, who maintains that accountability and care are not, and should not be, merely ethical abstractions but the result of what she calls "having truck with each other." This involves touch, regard, looking back, and "becoming with" as key modes of knowing, all of which make us responsible in "unpredictable ways for which worlds take shape" (2008, 36). Rather than setting aside ethical and ontological dilemmas in favor of some ideal normative principles, Haraway proposes "staying with the trouble" as a way of paying due attention to the messy realities of human-animal entanglements. In a similar vein, Annemarie Mol insists that in the ethics of care, "principles are rarely productive" (Mol, Moser, and Pols 2010, 13) and suggests instead a notion of care as embodied practices that "demand attuned attentiveness and adaptive tinkering" (15). Tim Ingold (2011) offers "wayfaring" as a template for inhabiting as well as knowing the world and suggests that we *follow what is going on,* tracing the multiple trails of becoming, wherever they lead" (14). I will return to these ideas later in the chapter, as I let them guide me through the choreographies of caring and killing as they unfold in ethnographic encounters. But first, let us turn to another location where sentience is being enacted, that of the journals and the laboratories of biology and veterinary science.

LOCATING SENTIENCE IN ANIMAL SCIENCES: NATURE, SALMON, AND THE ABSENCE OF NEOCORTEX

While the philosophical debate deals with "animals in general," biology and veterinary science engage animals in their specificity. This specificity is usually also species-specific and highly attentive to the capacities rendered possible in the bodies of different animals, especially as they are enacted through scientific experiments. These bodies are sometimes made to represent a particular fish species, such as steelhead or Atlantic salmon; at other times they represent fish in general, supporting the case for salmon sentience. Hence, the relevant debate is not about animal sentience in general but about fish—or salmon—sentience in particular. Bentham's "Can they suffer?" becomes a question of whether specific types of fish, such as salmon, are literally able to feel pain. Salmon sentience (or the lack of it) is thus located in the fish's body or, more precisely, in its physiological and neurological anatomy.

One of the most widely cited articles on this topic is a review by biologist James Rose (2002), which famously concluded that fish don't feel pain. Based on the arguments that (a) responses to "noxious stimuli" are separate from psychological experience of pain, (b) an awareness of pain and fear depends on specific functions of the cerebral cortex, and (c) fish lack these essential brain regions, the article concludes that the experience of pain and fear is unlikely for fish. According to the argument, sentience is located in the neocortex, an organ that fish do not possess. This conclusion has been challenged by a number of scholars who study neurological and biochemical receptors as well as fish behavior. The counterargument runs something like this: Even though fish lack the neocortex, which is key to the subjective experience of pain in humans, "the same job can be done in different parts of the brain in different kinds of animals" (Huntingford et al. 2006, 342). For these animals (and fish would be among them), "higher consciousness" associated with human evolution of the neocortex is *not* a necessary precursor for experiencing the adverse states that we humans associate with pain. Hence Huntingford et al. concluded that "taken together, these findings suggest that fish have the sense organs and the sensory processing systems required to perceive harmful stimuli and, probably, the central nervous systems necessary to experience at least some of the adverse states that we associate with pain in mammals" (342; see also Chandroo, Duncan, and Moccia 2004; and Lund et al. 2007, 112–13).

Alongside this debate and informing it are studies of fish cognition, which essentially ask about the sophistication of cognitive and behavioral processes in fish, evidence of which would support the argument that fish suffering is a real possibility. Research questions concern, for example, whether fish can recognize each other, whether they cooperate, how they learn, whether they can learn by observing each other, and whether they have spatial memory. A variety of fish species have been enrolled in such experiments, and because the answer to the research questions posed above is often yes, this is taken to support a more sophisticated set of cognitive skills in fish than was previously assumed and, hence, to support the argument that fish are likely to experience pain or fear even without a neocortex (Bshary, Wickler, and Fricke 2002; and Huntingford et al. 2006).[4] According to such arguments, sentience is not located in a specific part of the brain but can be inferred from studies of intraspecies interaction; thus, they expand the notion of what sentience is and where it may be found. However, and in line with the conventions of natural science, they tend to leave out human-animal relations.

When biologists ask about sentience in relation to specific fish species, such as salmon, they enact, through sampling strategies, a particular relation between the fish in the lab and fish bodies in general, making the former subordinate to the latter: the individual bodies of fish in the lab "matter" only to the extent that they can be made to enact a valid representation of the universe of salmon. But another way of looking at this is to hold that no animal suffers "in general": any instance of pain is necessarily specific, occurring in a particular body, at a particular time, and in a certain moment in that animal's path of becoming. As such it resides in relational practices, including those in the lab. In this perspective, for pain to actually *matter* would require not only that it be recognized by humans in the abstract as a potential generic capacity of a certain species but also that it be recognized and dealt with in the messy reality of coexistence, where simple solutions are rarely available. Pain, then, would be about the ability to respond, which requires not only the sensibilities, but also the *devices,* material and otherwise, that make possible a capacity to respond.[5] As Haraway (2008) notes, "Mattering is always inside connections that demand and enable response, not bare calculation or ranking. Response, of course, grows with the capacity to respond, that is, responsibility. Such a capacity can be shaped only in and for multidirectional relationships, in which always more than one responsive entity is in the process of becoming" (70–71). Thus, the legal frameworks and their scientific and philosophical underpinnings constitute only *one set of elements* in a farmed salmon's journey through life and death. Other elements include particular technologies, steel and plastic, knives and gloves and human hands, ice, water, and electrical voltage.

I approach sentience as something that is nurtured in the day-to-day practices of relating, across species boundaries in materially heterogeneous settings. The moral sensibility is, as Haraway (2008) puts it, "ruthlessly mundane" and constituted in the mutual ability to "respond" (75). Thus a choreography of caring and killing must approach sentience not only as a property of salmon but as a potential affordance of specific sociomaterial assemblages[6]. Sentience becomes, then, *an aspect of specific configurations* in "the articulation of bodies to other bodies" (Haraway 2008, 84), which may or may not be facilitated by the specific choreographies through which death is done.

Salmon are particularly interesting not only because they are newcomers to the regulatory frameworks of animal welfare but also because they remain *fish*. They are cold. They live in water. They are mostly out of sight. They are silent. Their eyes have no expressions that humans can recognize. Their body language is difficult for us to interpret. This limits the cues to which we humans can respond and thus makes the sharing of suffering a less likely, or perhaps a less dominant, aspect of the human-animal relation. Thus, if we follow Haraway's suggestion that the ability to share other animals' pain nonmimetically (and thus our capacity to respond) is a necessity if we are to act responsibly, we need to ask what exactly that might mean when it comes to salmon.

How can we—humans—take responsibility for the well-being of farmed fish? What are the practices through which we are enabled to know, or sense, fish as sentient beings? How do salmon "talk back"? How do we know anything at all, standing on a platform at a salmon farm, where most fish are always out of sight?

In order to address these questions, I explore the different ways in which death is being done on the salmon farm. Drawing attention to death is a way of highlighting the silent counterpart of life and growth, and of telling a story rarely told. But it also allows us to explore a terrain of human-animal relations, which are fraught with paradox, and within which are constantly working out what it means to be human and what it means to be fish. The idea is that attention to different choreographies of death and killing can illuminate the fragile achievements of life, and that this in turn speaks to the politics and practices of caring, eating, and living. Living well, eating well: these experiences are common to all of us, and areas where care and compassion resonate deeply with personal life experiences, which are often acquired at a young age. Sometimes such memories direct our gaze and motivate our response in ways we can neither escape nor properly account for.

I will introduce the ethnographic section by sharing one such memory that has informed my own approach to fish and sentience: I am in a small boat early in the morning, with my dad and my older brother. We are at Hardangervidda, a mountain plateau in southern Norway, where on summer holiday a nuclear family from an Oslo suburb briefly returns to the subsistence practices of its imagined ancestral past. A week at these remote lakes is a week of trout fishing, less for fun than for food. Our cabin is twenty kilometers from the nearest road, so we need to carry in everything we need for the whole week. Fishing takes up a great part of the days. Drift nets are left in the lake overnight in order to ensure a steady supply of trout. Every morning we fetch in the nets, and while my father pulls them up, and my brother steers the boat, my job is to remove the fish from the messy heap of nets and fish at the bottom of the boat, then place them in a bucket. A few are dead, but most of them are not, and I am careful not to hurt them as I gently disentangle the fine nylon netting from behind their fins and jawbones. Their teeth are tiny and sharp. When the job is done, I enjoy looking after them, so I scoop up water from outside the boat, just enough to cover their half-dead bodies. This goes on for a few days, until one day my dad intervenes. With few words, he lets me know that my concern serves the fish better if I put an end to their suffering. He shows me how to hold each fish, head facing forward, my thumb against its neck, and how much strength I need to hit it just right against the inside of the boat. I learn to notice the tension, like a sudden cramp, that tells me that I hit it right. I feel the moment of death in the palm of my hand. And I learn, somewhat reluctantly at the age of ten, that killing and caring can be done in a single quick movement.

CHOREOGRAPHIES OF DEATH: KILLING AND CARING

Death comes in different forms in different places. For farmed salmon, whose lives are all about putting on weight, most deaths are carefully timed, or scheduled, as a final step towards their realization as market value, as a food commodity. Scheduled deaths take place in specially designed salmon slaughterhouses, transforming "animal to edible" (Vialles 1994); this is how most salmon die. The scheduled death (and the subsequent cleaning and processing) is a "rite of passage," but also the moment when value is revealed, in the form of a tag that specifies its weight, buyer, and transit destination (Gothenburg, Amsterdam, or Oslo) en route to its final destination in Japan, China, Germany, or France (see chapter 4, " A Global Bulk Commodity").

Some forms of death are not scheduled and serve no particular human purpose. Rather than turning animal to edible, they turn animal to "dead fish," or *daufisk* in the Norwegian vernacular (see chapter 3). A *daufisk* must be removed; it is "dirt," distinctly nonedible, sometimes a disgusting slurry of bones and decaying flesh. Once removed, it will be ground up, mixed with formic acid, and transformed into feed for mink. If the scheduled death is a point on the salmon's passage from animal to edible and thus a confirmation of the ontological status of salmon as human food, the unscheduled death ends that passage once and for all.

Another type of death takes place in rivers in summer, when salmon run upstream to spawn. These are the deaths that happen when salmon are caught by anglers; they will not be dealt with ethnographically here, but see Nordeide (2012) and Ween (2012). These deaths are the outcome of a purposeful "hunt" for an animal that is thought to act and respond, in a game that distributes subjectivity and agency fairly equally between hunter and prey. There is also the occasional "death" of particular rivers after they have been treated with the chemical Rotenone, which kills a salmonid parasite, *Gyrodactylus salaris,* but also 'everything else' in the river until, it is hoped, the salmon return. Sometimes death comes to a particular genetic strain of Atlantic salmon, such as the salmon from the river Vosso, which are threatened or possibly becoming extinct (see chapter 7). Such deaths point to the absence of life or the failure of the cycle of spawning and migration to carry on, thus leading to irreversible losses of biodiversity. But as the Rotenone example indicates, particular "deaths" can sometimes be a last resort to restore a particular form of "life." Hence life and death are, even here, closely intertwined.

But in this chapter, we will remain with the farmed salmon. Let us return to Vidarøy, the grow-out site in the middle of the Hardangerfjord, where about 600,000 salmon have spent a year already.

REMOVING *DAUFISK:* NOTES FROM THE UNDERTAKER

I pick up a pair of blue rubber gloves and a pocket knife and place them in the wheelbarrow with the bucket that I rinsed out yesterday. I forget the paper form, rush back into the office to pick it up, and place it in the wheelbarrow next to the bucket. The form is attached to a metal plate and makes a rattling sound as I push the wheelbarrow about 125 meters to the far end of the metal ramp, and cage 10. It is time for the daily removal of dead fish.

Removing dead fish is a daily routine on the salmon farm. Recall from chapter 3 how the caretakers are also the "undertakers," as dead fish must be removed from the living. Hence, death becomes a very visible feature of the salmon farming enterprise.

I have turned the switch in the basement that turns on the compressor, so that air is pumped through a pipe that runs along the metal walkway. Now I connect the pipe to another that is connected with the "air lift" at the bottom of the cage. Soon I hear a hissing sound, and the big translucent tube on the surface begins to move around like a sea monster as water is blown up to the surface, until it suddenly pours out of the end of the big tube and into the blue plastic container on the ramp. I jump to the side to avoid the sudden shower, and I let it run for a couple of minutes, watching attentively as fish are washed up by the stream of water in the pipe. I am counting the "morning catch": one, two, then two small ones—and another big one, barely alive. After a while the dead fish stop coming up, so I turn off the pressure, disconnect the pipe, put on my gloves, pick up each fish by the tail, and drop it in the wheelbarrow. I cut the throat of the big one. Then I pick up the paper form with the pencil attached and scribble "5" next to the cage number, "10," and today's date.

Until they die, most grown salmon remain out of sight. Death makes farmed salmon visible to their human companions. This is how we see them. But death is also being counted. The number of dead fish is carefully inscribed on a sheet of paper and transferred to the electronic spreadsheets that aggregate the daily number of dead fish into the weekly and monthly reports of fish inventory. In this report the *daufisk* is translated as "economic loss." Hence, attending to the *daufisk* is more than a hygienic necessity: the inscriptions also help make the salmon assemblage visible in a managerial sense so that it can be more precisely monitored as a dynamic economic entity (see also chapters 3 and 4).

Doing the *daufisk* rounds every morning is one way of paying attention, of making the salmon "speak back." Feeding is another, and perhaps the most important mode of knowing salmon across the water surface (see chapter 3; see also Lien and Law 2011, Law and Lien 2014). Such encounters confirm to the farmhands that things are OK, alert them to potential problems, and serve in a small way to cultivate a human sensibility that is not purely anthropomorphic. Haraway (2008) calls for a robust nonanthropomorphic sensibility that is accountable to irreducible difference (90). On salmon farms, such sensibilities are cultivated in a fairly instrumental and indirect manner and revolve around ways of making the salmon "speak back" (Haraway 2008). Checking the feeding and collecting the dead fish are just two mundane examples of how a sensitivity to animal welfare is being performed—how sentience is being enacted—even when it is not being explicitly spoken about.

INVISIBLE DEATHS: NOTES FROM THE VACCINATION HUT

Not all fish that die appear on the surface as *daufisk* on the morning rounds. Some are deliberately led down a different path, one that separates them from the highway from parr to food. We became aware of this possibility of diversion during a wet, cold week in late autumn, when we took part in vaccination at the smolt production site. Vaccination happens a few weeks after the parr arrive and prepares them for a crowded life in tanks and pens. Vaccination is semi-automated. The fish are pumped from a nearby tank into a bath, where they are anesthetized. The injection is then done by a machine, which is fed with a steady flow of parr from a conveyor belt.

This is the first time that the young parr are handled as individuals and the first time that the smolt production manager has a chance to count them. The machine counts the fish and provides a number at the end of the day, which helps the manager decide how much to feed each tank during the following months.

But the machine also sorts the fish according to size. Fish must be at least 11 centimeters long in order for the needle to inject the belly as it is supposed to. Smaller fish are automatically detected on the conveyor belt just before they reach the needle and are washed into a groove that separates them from the rest and that leads to a pipe that flushes into one of the tanks outside. When we looked for them the next day, we found them in tank 15.[7] The tank was half full, there was no feeding system, and it held only a few handfuls of smallish fish—next to nothing in terms of numbers, a tiny minority in the city of fish. Why were they not being allowed to grow any bigger?

"No point," said the manager. "These are already having difficulties. If they aren't able to feed and grow during the first few weeks, they aren't likely to ever catch up." In the words of the manager, they are *tapere* ("losers")—made invisible through the ordering of the assembly line.

Enacting healthy salmon involves various practices of separation; the practices of ordering necessarily produce a shadowy hinterland of the other: those too small to be vaccinated are too different to be placed on the trajectory of further growth (see also Law and Lien 2014). Several things are going on here. First, this is an instance of standardization of each batch, or tank, as a collective whole. Being of equal size is seen as desirable for a number of reasons, including fish welfare (see chapter 4, "Becoming Biomass: Practices of Weighing and Counting"). Second, there is the anticipation of failure to thrive, attributed through the labeling of fish as losers. Third, this is an instance of care being done. Ideally, the vaccination needle is injected "immediately anterio-laterally to the pelvic fins, at an angle of about 25 degrees, so that the vaccine enters the peritoneal cavity with a minimum risk of damaging the underlying organs" (Stead and Laird 2002, 378). The machine is designed to vaccinate fish of fairly consistent size: it is programmed to place the needle at a certain distance from the nose of the fish. If it isn't separated out, a smaller fish will almost inevitably get a misplaced injection, with subsequent organ damage.

What happened next? A couple of days later, we checked on tank 15, but the smaller fish were gone. Asphyxiated, we are told, and ground up with the rest of the *daufisk*.

WHEN DEATH IS UNEXPECTED: NOTES FROM THE EMERGENCY TEAM

Sometimes death comes unexpectedly. John, my daughter, Eira, and I had arrived at the smolt production site on a January morning in 2010, in the middle of an exceptionally cold winter.[8] Hard snow covered the ground, and the river that supplies the tanks with running water was covered in thick ice. Temperatures around −15°C had persisted for weeks. The salmon farm is located in a region known for its relatively mild winters. For the nearly thirty years that this site had been in operation, there had been no need to invest in a water heater—until now. The children embraced this unexpected chance to go sledding every day after school, but the parents who worked with the smolts were in no position to enjoy it.

Within weeks, several hundred thousand healthy young fish, an excellent cohort scheduled to be smolts in spring, had become slow, drowsy schools of fish on the verge of dying.[9] Ice formed on the water surface and caught the fish as they gradually lost their sense of direction and came floating to the surface, belly up, drifting with the current. Feeding was hardly necessary. The *daufisk* round, which normally took less than an hour in the morning, now occupied the greater part of the day. Fingers froze and knees and arms got sore as we knelt on the side of the eight outdoor tanks, filling up one bucket after another and emptying them into a container nearby.

We counted the dead fish by the bucket and saw the number of fish in the tanks decrease day by day.[10] Every morning came with the hope of warmer weather, every afternoon with a deeper sense of loss. Exactly why the salmon died was not entirely clear. The vet was called and arrived in the afternoon, having been delayed after her car skidded off the icy mountain road. But she had little to offer except the consolation that at least we were doing the best we could.

"We know very little *(vi vet lite),*" she said. "We have done all the tests we can think of: water quality in the tanks and in the supply water, pH, aluminum, and iron. We have examined the gills and sent freeze-dried gill samples to a laboratory.[11] So far everything looks fine, and none of these results explain anything. Most likely we will conclude that what we see here is nothing but the result of the cold."

As days went by I learned to distinguish the nearly dead from the not-yet-dead. When in doubt, I picked them up, held them a few seconds in my gloved hand, and if I sensed any kind of movement, I threw them back in the tank. Otherwise, I dropped them into the bucket. As the bucket filled, my hand grew accustomed to the feeling of death, and my thoughts started to drift. Death became routine. But every once in a while, when a surface that we had cleared just an hour earlier was scattered again with white bellies, my heart sank. I felt heavy and lethargic.[12]

Does it matter that fish die? From a managerial perspective, each salmon that dies on the farm is in theory an economic loss. In practice, however, it all depends. Is there a good supply of smolts this season? Is there perhaps even an oversupply? In that case, the loss matters less. For the company as a whole, which operates several smolt production sites, the loss was not as dramatic as it appeared to us. There happened to be a surplus of smolts in the region that year and the lost ones would be replaced.

From a farmhand's perspective, it is different. The lunch conversations, which had been filled with laughter in October, were quieter and more somber, and laced with a black humor. Whenever it looked really bad, Kristin would shake her head and murmur: "*Det er deprimerande*" (It is depressing). There were rumors of workers who didn't sleep at night.

Returning to Oslo, after a week's work, I felt cold all the way through, and when I closed my eyes at night, images of dead smolts, floating belly up in the icy water, came to mind. Reflecting on our experiences, this is what I wrote in an e-mail to John:

> We got back fine, and Eira and I both crashed . . . both exhausted. And I have been cold, as if I need to thaw from the inside; it has taken a couple of days, with wool blankets and a fire going. Last night I couldn't sleep, I felt a strange intense sadness, and when allowing that feeling, the image of white belly-up fish flashed inside, some dying, some already frozen stiff, next to each other around the green plastic, and filling up that daufisk cylinder, so tight sometimes that I couldn't even squeeze the net inside, and had to start removing them by hand. It still makes me shiver, in a way it didn't when I was there. It is the visual image that is most disturbing.
>
> . . . This is just a private note, but at the same time, a form of data. I've killed fish regularly since childhood, and am not normally that sentimental about this. I accept the vet's claim that they probably don't feel pain. I think there is something about

death, and the liminal phase between life and death, and the physical manifestation of that phase, the way it feels between my hands, its weight, that gets me. The sheer numbers. The way it never stopped. Together with sore knees and the cold that went with it. All of this left a mark. Like a bad dream you want to shut off, but it keeps coming back.

I think about the word "depressing" that they all used. The only word they used. . . . I wonder what sort of effort it takes to endure this day after day. And the sadness that was there, underneath the gallows humor.

Uncertain about what to do with this piece of material, I decided to share some excerpts from my fieldnotes with Kristin. Another year had gone by, summer was approaching yet again, and a water heater had been installed in the meantime. There would not be another tragedy like the one we had witnessed. I had given her a printout of my ethnographic description the day before. "How did it feel?" I asked her as we shared a cup of coffee at her house. "You captured it well," she said. "One of the things that made it difficult was that we felt so helpless. You are *there* all the time, but you *still* fail, and there is nothing you can do to make it better. Sometimes it felt as if nobody really saw us, the way we struggled. But you did and you shared it. Thank you for that."

BECOMING FOOD: NOTES FROM THE SLAUGHTERHOUSE

We have arrived with the fish. One hundred and twenty metric tons of salmon have just traveled sixty-six nautical miles through the night, in two tanks submerged under the lower deck on a brand-new well boat that serves the industry in the counties of Rogaland and Hordaland. A crew of six and two ethnographers accompanies them on their final journey as lively flesh, which will take them to the salmon slaughterhouse and processing plant. Now the salmon are being flushed through a pipe from the well boat to the slaughterhouse dis-assembly line, which is on an elevated ramp in a large room overlooking the fjord.

From the narrow walkway on top of that ramp, we can see the entire hall and three men down below, busy bleeding the fish as they arrive, having been stunned just a few seconds earlier. As I stand on that ramp, I can feel the movement of their big bodies as they shoot out of the horizontal pipe and onto the conveyor belt that takes them slowly down the path towards the brand new electrical stunner. At this point they flap around like crazy, and the whole structure moves and shivers underneath our feet. The stunner is a metal box, placed on top of the conveyor belt. Stunning is automatic and adjustable; it is done a couple of times on each fish as groups of six pass through.

A supervisor comes up to us and says that the fish he has received today are really good, very strong and lively, so they need a strong electrical shock; in other words, they need to move slowly through the machine. For fish that are calmer, a shorter electrical shock will do, and they can go through more quickly. As he explains this, he opens the lid to show us how the stunner works, and for a few seconds we look down on fish held by metal claws and connected by thin cables. But when the lid opens, the electricity shuts off automatically, as a security measure for the workers. Soon we hear a call from down below, from the men where they are doing the bleeding. In front of them fish are piling up, flapping about and moving. Clearly, they have not been properly stunned and it is our fault. The boss quickly closes the lid and the stunning starts again.

The bleeding is what actually kills the fish; the stunning only renders them unconscious. The stunning machinery was put in place the previous summer in anticipation of new regulations for animal slaughter, which came into force in Norway on 1 July 2012. Previously, a carbon dioxide bath was used to render the fish insensible. But scientific reports by the European Food Safety Authority (EFSA 2004, 2009) concluded that carbon dioxide causes a strong adverse reaction and does not reliably result in unconsciousness; thus salmon risk being bled or eviscerated while they are still conscious (see Mejdell et al. 2010, 83).

Once they are bled, the salmon enter a slow rotating wheel to be flushed with cold water, which cools them down for about an hour before they are gutted and processed into whole fish sealed in Styrofoam boxes and loaded onto trailers waiting outside. Their journey through the slaughtering and processing hall lasts no more than two and a half hours. More than 30,000 salmon, each weighing 4–5 kilos will pass through the dis-assembly line today, in an operation that secures an income for local workers, some of whom are refugees. The workers' homelands seem as geographically dispersed as the destinations of the boxes of salmon that they prepare: their countries of origin include Morocco, Iraq, Hungary, Lithuania, Japan, Kuwait, and France.[13]

Having spent the entire day at the slaughterhouse without any mention of animal welfare, I decided to bring this up with the manager. Arne was eager to talk, and emphasizing that what he said was a personal view and that he didn't really know how the salmon actually respond (to learn more about that, he suggested, I ought to speak to a vet), he admitted that he didn't feel very strongly about the fact that fish were stunned or occasionally not stunned. He quickly added that of course they do all of this as well as possible, with the least possible discomfort for the fish. And the machine, adapted to the new regulations, is designed specifically for this purpose. But emotionally he didn't feel very strongly about it, and he had a theory about why this is so.

Arne is also a recreational deer hunter, like many others in this region, and when it comes to deer, his feelings are very different. He thinks that, for him at least, it has to do with the fish being cold: When you reach inside the belly of a deer, he says, and feel the heart still pounding, it does something to you. You really do not want to cause that animal to suffer unnecessarily.

He says that there is a constant effort in hunting to avoid injuring the deer, and this is why much emphasis is placed on aiming carefully. But when a similar thing happens to salmon, it doesn't affect him in the same way. He thinks this is because the deer is warm, like his own body, while the salmon feels cold and so in a way as if it were already dead. He presents this to me as his private theory but also as a puzzle that he has reflected upon.[14] That being said, he is completely happy to follow the current requirements about stunning, and has no strong views about it.

This conversation reflects one I had earlier with the manager of another marine site, where the fish had spent the previous two years. As we balance on the side of the pen, watching salmon being sucked up into the well boat, he tells me that he has helped out people raising Scottish Highland cattle. When the animals were slaughtered on farms, the meat was excellent quality, very tender. But when the animals were transported by boat and truck to the centralized slaughterhouse, the meat was like shoe leather. He uses this story to underscore a more general concern about not stressing the salmon unnecessarily. He then quickly draws on another story, from deer hunting: his father always used to say that the best deer meat is what you get when you shoot an animal that has not yet noticed you. It is the same thing, he says.

As we look down in the net that gradually pulls the salmon closer together, we see fish that move quickly, before shooting off in a different direction, splashing as they turn. It seems that in spite of all legal efforts to prevent unnecessary suffering in farmed salmon, some moments of stress are unavoidable.

BECOMING ANIMAL: ENACTING SENTIENCE THROUGH LEGISLATION

Sentience can unfold in a number of different ways, through different sets of practices and through different devices. I have drawn attention to how sentience is situated, relational, and anchored in day-to-day practices on the salmon farm. Legislative measures and the paper trails they produce tell of another way in which sentience unfolds. Such texts are carefully crafted justifications, based on assumptions that certain qualities, such as animal sentience, are universal. But their scope for jurisdiction remains territorially bounded and politically situated. Let us turn to Norway and the EU and the recent changes in the regulations of animal welfare concerning farmed salmon.

It is often assumed that legal structures somehow correspond to natural or social facts, but recent studies of law have taken a different approach, emphasizing that what appears as a fact is also sustained by social practices, including legal practices (Pottage 2004, 2–3; Asdal 2012). Hence, just as legal structures sustain ontological distinctions between persons and things, legal techniques can also enact—or undermine—other categorical distinctions, such as that between animals and fish.

The first comprehensive "cruelty to animals act" was passed in England in 1876 (Lund et al. 2007), and mistreatment of animals was established as a criminal act in Norway in 1902 (Asdal 2012). About a century later, the European Union placed animal sentience on its legislative agenda. According to Turner (2006), this happened in 1997, when an annex to the legally binding Treaty of Amsterdam recognized that animals were "sentient beings" and required the EU member states to "pay full regard to the welfare requirements of animals." In 2004, the European Food Safety Authority (EFSA) had issued a scientifically based opinion concerning transport and stunning and killing of farmed fish, and concluded that many existing commercial killing methods expose fish to substantial suffering over a prolonged period of time. These documents make suffering a feature of farmed fish, and thus justify a space for fish in the legal context of animal welfare legislation. They do so by making the argument that in relation to legal responsibilities in fish farming, "fish are animals too." As such, these documents do "ontological politics" (Mol 1999); they interfere in the order of things, the basic categories of being, and in doing so they become a source of justification for subsequent legislation.

In 2005, the Council of Europe (CoE)'s Standing Committee for the Protection of Animals Kept for Farming Purposes passed new recommendations for farmed fish (Lund et al. 2007). These recommendations, which were passed under the Convention for the Protection of Animals for Farming Purposes, ratified in 2005, and became effective as of 5 June 2006, state that "in light of established experience and scientific knowledge about the biological needs of fish, methods of husbandry at present in commercial use may fail to meet all their needs and hence result in poor welfare" (article 7) and that "if the fish are to be killed, this shall be done humanely" (article 5.3; Council of Europe 2005).

Legally binding regulations appeared in 2009, when the Council of the European Union passed the Regulation on the Protection of Animals at the Time of Killing. Article 3.1 states: "Animals shall be spared any avoidable pain, distress or suffering during their killing and related operations." However, specific standards listing what exactly that means when it comes to fish have not yet been passed.[15] In the meantime, a similar move to include farmed fish in animal welfare legislation took place in Norway, which is not an EU member state. A new Norwegian animal welfare law *(Dyrevelferdsloven)* came into effect in 2010,[16] replacing the 1974 law on animal protection *(Dyrevernloven)*. In 2008, more specific regulations of slaughterhouses for "aquaculture animals" *(akvakulturdyr)* had already been proposed; they banned the use of carbon dioxide and thus implied major changes in slaughterhouse technologies. The replacement of the term *aquaculture fish* with *aquaculture animals* can be seen as another example of ontological politics, placing the farmed salmon firmly within the category of farm animals. These new Norwegian slaughter regulations treat farmed fish quite explicitly, requiring that fish be stunned before they are bled. They define electrical stunning and percussive stunning as the only acceptable methods, on the grounds that these cause less pain than the alternatives that have been commonly used (for example, carbon dioxide and clove oil). In practical terms, this means that farmed fish are killed in the same way as terrestrial farm animals. The regulations regarding slaughter were not fully enforced in Norway until 1 July 2012; they are more specific than the current regulations in the European Union in that they specify the precise slaughtering methods that can be used and ban methods that are common in many other countries.[17]

In the United States animal welfare is regulated through the Animal Welfare Act of 1966. According to the amended Animal Welfare Act in 2009, the term *animal* refers only to warm-blooded animals; although it excludes birds and mice, among others, it does not mention fish.[18] The same definition is applied in more specific animal welfare regulations; hence, it seems reasonable to conclude that, although some guidelines call for "respectful treatment of wild fishes in field research,"[19] farmed fish are not legally protected from unnecessary suffering in the United States the way they are in Europe (with general protection from avoidable pain) and in Norway (with additional specific regulations regarding slaughtering methods).

Legal regulations can be seen as one of many moves enacting salmon as sentient beings and society as a moral collective.[20] In the practical work of governance, they also serve as checklists. Thus, when current Norwegian animal welfare regulations state, for example, that "fish shall be protected from unnecessary stress, pain and suffering at the time of slaughter" (*fisk skal vernes mot unngåelig smerte, lidelse og frykt ved avliving og aktiviteter I forbindelse med det*) (Article 10) and that "it is illegal to anesthetize fish by using gas, including CO_2, or any other medium that blocks the oxygen uptake (Article 14), this establishes a particular standard for slaughter and gives the veterinary inspectors some criteria to follow.[21] But in the messy reality of salmon assemblages, such legal guidelines appear both too idealistic and at the same time insufficient, as no checklist can ever completely remove the moral and practical dilemmas involved in raising farm animals. Let us turn to ethnography again.

ACCOUNTABILITY AND CARE: NOTES FROM A MANDATORY WELFARE COURSE

"It can always be done better." That is the idea behind the new animal welfare regulations in Norway, which require that as of 2010 all fish-farm workers shall attend fish welfare courses on a regular basis. We are gathered in the light, spacious attic of the head office building. It is early February 2012, the second day of this two-day intensive welfare course; Maria, a young veterinarian, is in charge, together with a senior operation manager. She facilitates these courses on a regular basis, typically a couple of times a year, as part of her job. The students—about twenty employees plus the anthropologist—come from different locations owned by the company. Some have spent the night in an apartment next door, and others have come by car and an early morning ferry to get here in time. The first day has covered a lot: animal rights, ethics, philosophy, cultural variations in human-animal relations, the five freedoms, fish biology, physiology, and the salmon's "natural needs," as well as the new legal regulatory framework. Today, we are focusing more directly on day-to-day practices. We have been divided into two groups. I am with the smolt production group, where I know about half the people, including a couple of senior managers. I spot a few recently hired employees, whom I recognize from when they were introduced at a Christmas party in December. Everybody has to do this course, so it is "back to school" for all, irrespective of position or previous training. It is another day of PowerPoint presentations, lectures, question-and-answer periods, and lots of coffee, snacks, and chatting in between.

In the morning session we hear about vaccination, the recommended length and diameter of syringes, water temperatures, oxygen levels, surveillance systems, pipes, and transport. It's enough to get my head spinning, but at the same time it is all very practical, with photos, some of which show people in the room. We are taught about procedures that they and I have already learned through practice, but now we relearn them with the added focus on *why* it is this way or that way, on thresholds, on matters related to fish welfare, and on how it is done differently in different locations. We are comparing notes from different sites as we discuss the pros and cons of various approaches. In the afternoon comes the dreaded group work session—our assignment: "Go through the complete production cycle in your workplace, identify welfare issues and improvement potential, and relate it to the current fish welfare regulations. Present a plan for improvements, or justify why it is good enough as it is" (my translation).

As the only professor and also the only woman in my group, I am immediately elected the secretary. We sit around talking for an hour or so, and based on what is being said, I make a tentative list of seven points to present to the entire group in the soon-to-follow plenary. One bullet point concerns the size of pipes that are used to transport smolts in and out of different tanks and how the two pipes hooked together need to be of *exactly* the same size. The reason is that if fish are flushed from a larger to a smaller pipe, the hydromechanical pressure inside the pipes inevitably creates a "traffic jam"—fish get stuck, and even if they all flush through eventually, it does not appear to be a comfortable place to be stuck. These and other details that I have never encountered in legal documents, or even thought about, now become the focus of some discussion, simply because someone brings them up.

Do fish feel pain? This question, which is still contested among biologists, somehow seems less relevant here. Or perhaps it is replaced by the more immediate issues of concern, such as, *How can I avoid hurting the fish unnecessarily?* In the end, one of our bullet points concerns the new shelves for fertilized roe, conveniently placed on top of each other like drawers in a walk-in closet. The ones they used before were much wider and placed about waist-high, causing a lot of repetitive back bending, and stiffness and back pain as a result.

"Is this *really* about fish welfare?" someone asks.

The group is undecided for a moment, until the manager of a smolt production site uses his authority to settle the issue: "Of course it is," he says. "If people smile, then the fish are happy too." Then he adds a final point about tidiness, because stumbling over rubbish causes frustration, and human irritation is bound to affect the fish one way or another.

This is, of course, not the first time we have talked about how to take good care of the fish. As indicated in previous chapters, concerns with fish welfare and fish health permeate most of the day-to-day practices and a good part of the trivial talk along the tanks and pens. Nevertheless, the welfare course is the first time that I have encountered a discursive space explicitly and almost exclusively focused on this issue. In this way, it offers a legitimate space for farm workers to articulate their own welfare concerns—that is, to verbally enact themselves as sentient human beings in relation to the fish. In doing so, they simultaneously enact the salmon as a sentient being, collectively, explicitly, and with the aim of taking responsibility, or becoming more "response-able" in the relationship. Rather than merely being a philosophical concern, sentience is enacted here as part of the messy and heterogeneous human-animal assemblage, one in which there are newcomers at both ends: salmon are "newcomers to the farm," and we who care for them are newcomers to this underwater world of farmed salmon. We do not know very much, but that does not stop us from making each other somewhat accountable.

Is this an example of what Donna Haraway (2008) calls nonmimetic caring? Perhaps we are collectively building, in a small but significant way, that "robust non-anthropomorphic sensibility that is accountable to irreducible difference" (90). Or perhaps we should rather think about this as a small part in a much larger move towards what Mol, Moser and Pols (2010) refer to as "good care," or "persistent tinkering in a world full of complex ambivalence and shifting tensions" (14). Perhaps the group session is a kind of wayfaring, or simply a practical and illustrative case, that shows when it comes to care, "qualification does not precede practices, but forms a part of them. The good is not something to pass a judgment on, in general terms and from the outside, but something to *do*, in practice, as care goes on" (13).

Another day, I meet with Maria again, and at the end of a conversation in which she explains some details about sea lice, pen treatments, and current regulatory measures, I ask her, *"Er fisk dyr?"* (Are fish animals?)

She pauses for a moment, and then replies, "They are covered by animal welfare legislation. That makes me, as a vet, their spokesperson. And I think that is a good thing."

ENACTING SENTIENT BEINGS

So what have we learned about sentience? Rather than offering a definition, I have offered an ethnographic journey to different sites in which salmon sentience somehow came about. We have seen that ontological choreographies enact salmon sentience through different sets of practices. We have seen how salmon sentience, in the context of a brand-new slaughterhouse, is enacted through welfare regulation and then translated into specific spatiotemporal choreographies of voltages and bleeding. Sentimentality is not needed. The fish is already stunned, unconscious. The potential pain is taken care of, as separations unfold: The moral dilemma is quite literally delegated to the machines, which do the care for us. From the salmon's point of view, it may be OK. Compassion is not always the best guide.

And there are other times when it is hard not to be compassionate. When healthy fish you care for freeze in front of your eyes, a sense of loss seems unavoidable. This does not necessarily help them at the moment. But the following year, a water heater was installed. The regulatory requirements incorporated the possibility of harsh winter weather. But the workers required it too. There won't be another crisis exactly like the one they went through.

In tank 15, fish became invisible: sorted away before they were even counted. Through a mechanical ordering device that also performed vaccination, the losers were silently flushed away; even before they died, it was as if they had never existed. Their numbers were ridiculously low in the city of fish. A utilitarian approach to animal welfare would render them invisible too: they were too few to worry about. But in this story, they serve to illustrate the subtler nuance of an ontological choreography that enacts farmed salmon as visible, sentient, and properly cared for. Fish may be food, it may be waste; we may be instrumental, we may be emotional. A machine may emerge as a sorting device issuing "death sentences" or as a material manifestation of welfare regulations that appear to render the day-to-day care of dying animals superfluous. It will never be perfect, and we are never innocent. But it can always be done better.

"Making it better" calls for a move from the realm of animal rights and legal texts to the messy, murky corners of slaughterhouses, grow-out sites, and smolt-production tanks, where killing and caring go together. These are complex assemblages where our mutual capacity to respond across the species barrier is a possibility, but one that is not always realized—sites where the farmed fish are currently, literally, being constructed.

Rather than locating salmon sentience in the realm of consumer activism and animal rights movements or in scientific laboratories, I have searched for sentience as it is enacted in human-animal relations as they are done in salmon farming. Domestication has not always, and does not necessarily, enact sentience as a relevant dimension. Canada, Chile, and the United States are examples of countries where, at least in the legal realm and for the time being, fish sentience plays a different and probably less prominent role in food production. And yet, as we have seen, the coming together of humans and salmon in a setting in which the former are quite literally responsible for the growth and well-being of the latter may allow an emergent nonanthropomorphic sensibility to unfold. It is fragile and uncertain, and it hardly prevents suffering. But it enacts salmon sentience and nurtures a capacity to respond.

I could have approached these cases with the aim of disentangling the logic of expansive corporate capitalism from the logic of care that many of the employees—who are also small-scale, part-time farmers—bring with them. I could have juxtaposed the standardization inherent in industrial aquaculture with the concerns of local small-scale fishermen. I could have juxtaposed technology and care or teased commodities apart from moral subjects. But rather than rehearsing such dualisms, I have tried to show, first, that even in highly commoditized food production sites such as this, sentience and care may unfold. And second, that in practice, care is chronically uncertain, it is noncoherent, and it is multiple. Caring for the individual and caring for the collective don't always go together. Technology and care go together rather well sometimes. And caring and killing can be done in the same move.

"Making it better" becomes, then, a precarious ontological choreography that is constantly inventing ways to make the salmon "speak back" as sentient beings. It also calls for an environment in which workers aren't muted or automated (like the salmon on the dis-assembly line) but rather invited to engage the tools and devices available to them to tap into our shared capacity to respond to other species, even when they remain almost always out of sight.

. . .

As I was working on this chapter, I told Arturo Escobar, a fellow anthropologist, about my current work and briefly stated that in legal terms, farmed salmon in Europe are now becoming sentient beings. Having spent most of his time in the Americas, he was quite surprised, and asked, "Why is it that salmon are now seen as sentient? Is it the result of a pressure from consumers? Does the animal rights movement have anything to do with it? Is it the result of recent biological research, or perhaps reflecting a move in philosophy?"[22]

Puzzled for a moment by how difficult it was for me to answer such a seemingly simple question, I quickly conceded that yes, all of these play a role. But none are responsible. Because in the messy modality of care, such principles, ethical or scientific, aren't all that productive. Acting responsibly in relation to salmon involves practical tinkering, which is part of the process of dealing with farmed animals. By the time I had thought this through, my interlocutor had gone. My delayed reply, then, is that salmon are becoming sentient because if they do suffer, they no longer suffer alone: they suffer *in our care*. They are becoming sentient because, or rather *as*, they are becoming domesticated.

7

BECOMING ALIEN

Back to the River

Two men walk up the slope up from the river. Each of them carries a salmon. The man in front has a black hat with feathers and several pins attached to the brim. Inside his jacket he wears a red-checked flannel shirt. The man behind him is more broadly built. He is the boatman, referred to as "rower"; the man in front is the fisherman. The rower places the salmon in front of us, while the fisherman walks to the side and places his salmon on the ground, a deliberate distance from the other one.

"This is *ufesk*![1] One can easily tell that it is farmed; it has 'Salmar' written on its head," says the fisherman, clearly disappointed.

The rower adds that he is glad he got that shit out of the river. The river owner,[2] who stands next to them, pulls out some brown envelopes and says they have to take scale samples, as NINA [the Norwegian Institute for Nature Research] has asked them to do. The rower pulls out a big knife, kneels down over the fish—the one he just referred to as "shit"—and scrapes off scales, which he then seals inside the brown envelope. Everyone takes a second look. They conclude that there is no doubt about this one: its dorsal fin is mostly gone. (Nordeide 2012, 81; my translation)

The scene is the river Namsen, on the Norwegian west coast, but it could have happened on almost any of the several hundred rivers in Norway where salmon angling is permitted during the summer season. Namsen is one of the many well-known rivers where temporary salmon permits can be purchased at auctions each fall.[3] While the fisherman in the story has purchased his permit to fish from the river owner, the rower is being paid by the day to guide the fisherman to the best spots for salmon angling. Together they enact the colonial asymmetries that have defined salmon river angling in Norway since the arrival of British lords in the mid-nineteenth century (Solhaug 1983). I have borrowed this story from anthropologist

Anita Nordeide (2012), who did fieldwork on the Namsen River in 2011. I cite her story because it captures a very common practice in which farmed and wild salmon are enacted and made separate in Norway today. Before the 1980s the main differentiation in Norwegian rivers was between trout and salmon. But following the growth of commercial salmon farming, differentiations have multiplied. Farmed salmon now inhabit the Norwegian coastline from south to north. The most important distinction today is that between "wild" and "farmed," as illustrated in the story above. Most farmed salmon are securely contained in cages and pens along the fjords and in inlets. But accidents do happen. A propeller may occasionally rip the net open, and a few thousand salmon may get out. These salmon are generally referred to as "escaped farmed salmon" and are now managed as "alien species" in their Norwegian rivers of origin (Lien and Law 2011).

How can it be that so-called "escaped farmed salmon," the progeny of native Atlantic salmon that were removed from these rivers some eight or nine generations earlier, have become "alien species" in their rivers of origin? What does it mean to be "wild" in these rivers? And how can we describe salmon that slip away, not only from their pens but from human ordering practices too? This chapter traces salmon back to their rivers of origin. By shifting the focus from salmon farms to salmon rivers, a broader story can be told: one not only about the damage to salmon habitats and losses of biodiversity but also about ongoing experiments with recovery and about dealing with messy underwater worlds.

A QUESTION OF SCALE: SALMON RETURNS VERSUS SALMON ESCAPEES

Since its emergence in the early 1970s, salmon farming has grown exponentially (see chapter 5). During the period 2010–2012, an estimated average of 250 million salmon smolts were placed in marine aquaculture sites along the Norwegian coast each year (Directorate of Fisheries 2013a).[4] Adding the ones that were already there (recall that Atlantic salmon spend 12 to 18 months in marine pens), the total number of farmed Atlantic salmon along the entire coast at the end of January 2012 was 345,201,000, or nearly 350 million fish. All of them are descended from Atlantic salmon sourced in the 1980s and 1990s from broodstock caught along the same coastline. They are the progeny of broodstock that were selectively bred in order to optimize hereditary characteristics that are desirable for farmed salmon, such as fast growth, sexual maturation, distribution of fat, and disease resistance (see chapter 3); however, they are still considered to be the same species as their wilder cousins. Only a tiny fraction of these farmed fish get out, but since the total number of fish is so large, the flow of escapees can be quite substantial, although it varies widely from year to year. Based on reports submitted by the industry and analyzed by the Directorate of Fisheries (2013b), the average number of Atlantic salmon escapees (*rømte laks*) for the period 2002–2012 is 440,000 escapees per year.[5] Many escaped farmed salmon are caught, and many will die, but some are likely to make it upstream, where a few may be able to spawn.

Incidentally, this estimated number of escaped farmed salmon (440,000) is almost exactly the same as the number of wild Atlantic salmon that return to Norwegian rivers to spawn every year, according to estimates from the Norwegian Environment Agency (2012). According to its figures, the annual salmon return in 2012 and previous years was between 400,000 and 600,000, which is less than half of the estimated return in the mid-1980s. Hence, even if only a fraction of the escaped farmed salmon actually spawns, they could still be a significant addition to the stock. Escaped farmed salmon are able to interbreed, and the offspring are a concern for both fishermen and biologists in terms of not only classification but also preserving biodiversity. The genetic effects of interbreeding depend on the number of escaped salmon and their spawning success as well as the degree of genetic differentiation between wild and domestic salmon (Taranger et al. 2011, 37–39). A study from 2013 showed significant changes in salmon populations resulting from interbreeding in five of the twenty rivers that were included in the study.[6] The main question that biologists ask, in addition to spawning success, concerns the respective fitness and survival rates of the offspring of wild and farmed salmon. Even though the answer is not entirely clear, there is sufficient reason to exercise caution.[7] Hence, Norwegian authorities now see escaped farmed salmon as one of the many current threats to the wild salmon population in Norway and one of the two main mechanisms by which salmon aquaculture may have a negative effect on the wild salmon populations.[8]

But the wild salmon did not evolve in isolation before the 1980s. Salmon fishermen and their organizations have been diverting salmon spawning journeys for generations, at least since the mid-nineteenth century (Treimo 2007).[9] Eager to enhance the salmon stock in local rivers, they mixed eggs and milt indiscriminately and distributed young salmon fry across watersheds, fjords, and mountain ranges. Such diversions complicate current restoration projects and cast some doubt on the historical depth of salmon's genetically evolved fitness in relation to particular rivers. So in the flow of seemingly pristine salmon rivers, like the Namsen and the Alta, there is also a long and shifting history of cospecies evolution, which has recently intensified. Salmon's recent emergence as a husbandry animal in its Norwegian fjords of origin makes it ambiguous: neither quite domesticated nor completely wild, it upsets the ordering binary of "nature" and "culture" that domestication otherwise sustains.

In what follows, I first give a brief account of the emergence of salmon as an "alien species" and how the coining of the term is also an instance of performing Norwegian nature.[10] The account is based on a controversy that unfolded on the World Wide Web in 2007 and shows how words perform both nature and salmon simultaneously. The sections that follow turn to river practices, focusing on a recent and rather successful salmon rescue project on the river Vosso, not far from the salmon farming region where we spent most of our time during fieldwork. The third section takes us to an unexpected meandering of salmon trajectories, while the fourth section explores multispecies ethnography beyond material practices in an effort to engage salmon differently, through a playful act of imagining and naming that which we humans can hardly ever see or touch or hold.

"BECOMING ALIEN"; OR, THE PERFORMANCE OF NATURE AS A PLACE WHERE HUMANS ARE NOT

On 31 May 2007, Norway's Institute for Marine Research placed an article on its website with the heading: "Escaped farmed salmon is not an alien species" (Rømt oppdrettslaks er ikke en fremmed art). The statement was a direct challenge to the Norwegian Biodiversity Information Centre (NBIC [Artsdatabanken]), a governmental agency responsible for monitoring biodiversity in Norway, and the inclusion of Atlantic salmon in its recent publication on alien species called the *Norwegian Black List* (Gederaas, Salvesen, and Viken 2007). The black list, first published 2007, was an attempt to name and order all alien species *(fremmede arter)* in Norway based on their threat to local biodiversity. Shortly after the article from the Institute for Marine Research appeared, which challenged the classification of escaped farmed salmon as an alien species, the NBIC issued a clarification explaining why farmed Atlantic salmon were included even though they were only seven generations removed from their ancestors upriver.

This brief controversy tells us that salmon is contested in Norway, and not only at the level of individual actors with different interests such as anglers and plant operators. Research institutions disagree too, and at stake is a very basic ontological settlement of what a farmed salmon is. Is it an alien species or is it not? Clearly (or so it seems, from the way the arguments are framed), it cannot be both alien and not alien at the same time. Salmon is enacted through practices that are at once material, discursive, and social. Elsewhere we have argued that performing salmon in Norway is a performing of nature (Lien and Law 2011). What makes this particular debate relevant here is the way in which it provides an ontological canvas for further enactments in rivers. Let us turn to the black list.

The *Norwegian Black List* lists a total of 2,483 alien species in Norway and analyzes the ecological risks posed by 217 of them. It applies the definitions of the World Conservation Union (IUCN), which defines alien species as "non-native, non-indigenous, foreign, exotic . . . a species, subspecies, or lower taxon occurring outside of its natural range (past or present) and dispersal potential *(i.e., outside the range it occupies naturally or could not occupy without direct or indirect introduction or care by humans)* and includes any part, gametes, or propagule of such species that might survive and subsequently reproduce"(Gederaas, Salvesen, and Viken 2007, 9; my emphasis). This definition is a scaling device; it makes a particular aspect of Norwegian nature legible to a global discourse on biodiversity. Usually, the distinction between alien and indigenous species rests upon notions of geographical spread. Alien species are quite literally "species out of place" (Lien and Davison 2010). But there is a twist: The geographical spread is not always a problem. Species can evade the "alien" label if their dispersal is not caused by humans—that is, if they were somehow able to spread without the aid of human beings. Indigeneity rests on this separation: presence without people; nature without society (Lien and Law 2011). The full list of criteria that define alien species in Norway is as follows:

a. Species intentionally released into the wild
b. Species escaped from captivity and breeding, or run wild from cultivation and commercial activity
c. Species arrived as stowaways during transportation or movement of animals, goods, and people
d. Species dispersed from wild populations in neighboring countries whose origin is due to a, b or c.
e. Species spread with the aid of human beings
f. Norwegian (indigenous) species spread to new parts of Norway by human activity
g. Improved, indigenous species spread in Norway. (Gederaas, Salvesen, and Viken 2007, 16–17)

Items b and g most concern us here. These are how farmed Atlantic salmon become classified as alien. These items, especially g, point to a nonterritorial form of invasion. In Norway, farmed Atlantic salmon are within their "natural range." So why are they alien? Why are they not "natural"? Challenged by the Norwegian Institute of Marine Research, the Norwegian Biodiversity Information Centre responded that it had decided to include "improved, indigenous species" under the category of "alien species" because such "alien genotypes may represent a serious environmental problem" and it is necessary to respond to "threats to biological diversity at all levels, including ecosystems, habitats, species, and genes." The NBIC also noted that domesticated farmed salmon have had their hereditary material changed through "artificial selection with the aim of creating fish with the best possible characteristics for being raised as food" (cited in Lien and Law 2011, 79). The word *artificial* is pivotal. There is always genetic selection, but industry selects for particular traits. As we have seen, growth is particularly important. Biologists know less about "natural selection" than they know about selective breeding, though it is unlikely that traits such as "rapid growth" or "flesh quality" add to adaptability in most riverine environments. Again, the difference has to do with human intervention—or its absence. Human activity may *move* species beyond their (nonhuman and therefore natural) geographical range. Or, as with salmon, it may intervene in ways that alter the *constitution* of a species. Either way, humans have intervened in what was a "natural" process, which means that it no longer counts as natural. In the present context this turns farmed salmon into a threat to the natural gene pool:

> Some indigenous species are domesticated and have had their genes altered by artificial selection. If such species escape or run wild, domesticated individuals may hybridize with individuals in the wild populations. The wild forms may thereby be supplied with genes that are poorly adapted to the natural conditions. Such hybrids can result in decreased survival of offspring and a generally poorer adaptation to natural conditions. Examples of this from Norway are the wild salmon *(Salmo salar)* and the arctic fox which can receive genes from farmed animals. Aquaculture in particular has a number of species belonging to this category. (Gederaas, Salvesen, and Viken 2007, 40)

Most biologists agreed at the time that salmon had become distinctly different as a result of domestication, and the significance of these differences was a matter for debate in journals of fish biology (Gross 1998, Huntingford 2004). But the division between domesticated Atlantic salmon on the one hand and wild salmon on the other was, as we have seen, enacted in other practices too. In the following section, I will trace this division through further enactments as they take place on the Vosso River, focusing first on practices of differentiation in the Vosso salmon rescue project.

THE VOSSO SALMON RESCUE PROJECT

Vosso used to be known for its large Atlantic salmon. Old black-and-white photographs of fishermen posing with large freshly caught salmon tell stories about local pride and have become a symbol of the natural splendor of western Norway. Today, they also serve to document what has been lost and to mobilize around the ongoing Vosso salmon rescue project.

Until the 1980s the river attracted salmon fishermen from near and far, who returned every summer to fish salmon at sites known for their good fishing. Visitors, both British and Norwegian, paid local farmers, who were also by definition river owners, to lease their stretch of the river for the entire season. Sometimes they even built houses by the riverside and brought their families, establishing bonds that lasted for generations and contributed significantly to the local farmers' income. Vosso can be difficult to fish, but its salmon were larger than elsewhere. Between 1949 and 1987 an estimated average of 1,150 salmon averaging 10 kilos each were caught in the river each year. But by the late 1980s the Vosso salmon was on the verge of extinction. Original salmon stock was then taken out of the river and placed in what is called a live gene bank, in a mountain lake (Eidfjord), and a moratorium on fishing was put in place in 1992. A rescue plan was later developed,[11] and we had the chance to follow its implementation during 2011.

The decline of Vosso salmon is understood by biologists to be the result of a wide variety of factors, some of which have been mitigated, others not. These include hydroelectric development, acid rain, road construction, and—more recently—salmon farming, through increased prevalence of sea lice in the fjord and through hybridization and competition during spawning (Barlaup 2008).

The rescue project rests on two pillars:[12]

1. Cultivation of genetically distinct roe (from the gene bank) and release of large quantities of roe, fry, and smolt at various locations in the river watershed (from the river through the fjord to the coast)
2. Measures to reduce the current threats to Vosso salmon

Unlike that in the United States (for example, in Alaska and along the Columbia River), where a large proportion of Pacific salmon is now sustained through hatcheries, Norwegian salmon fisheries' management rarely involves substantial cultivation measures (Ween and Colombi 2013). Rather, a "wild salmon" in Norway is ideally autonomous, in the sense that it is able to complete the cycle "from egg to egg" without human intervention.[13] This involves hatching in a river, smoltifying after a year or two, swimming towards the ocean (often through long, narrow fjords), feeding and growing somewhere in the North Atlantic, and returning again to spawn successfully in the river of origin. Hence, the cultivation measures initiated by the Vosso salmon rescue project are seen as temporary, with the goal of promoting a self-reproducing, viable, and harvestable salmon stock after years of decline.[14]

The first release of eggs, fry, and smolts from the Vosso salmon rescue project took place in 2009; it was continuing in 2014. In 2011, there had already been a significant increase of salmon returns. The trend continued in 2012, triggering optimism and renewed enthusiasm among everyone involved.

But what does it mean that the salmon return? With a moratorium on salmon fishing, how do the salmon appear? Until the 1980s, encounters facilitated by a wet fly and a fishing rod or a dip net or drift net were the most important ways that salmon became visible to people; hence "catch statistics" were the most reliable source of knowledge of the salmon during this period.

With a moratorium on fishing, such methods are no longer available. Now a combination of other techniques is used to map various sections of the salmon river watershed and to measure salmon mobility. These techniques are all designed around the assumption that salmon are fundamentally mobile; often they also take into consideration the topographic configuration of a river. For example, the fact that all salmon need to move back up the river to spawn offers some possibilities for counting.

One technique involves the use of underwater cameras. A camera can be placed at a specific site in the river where salmon are expected to pass; linked to a computer, it allows the registration (manual or electronic) of the number of salmon bodies passing a certain point during a defined time period. High counts are promising, but not enough, because a camera does not reveal any details about the returning salmon.

Before the 1980s, such details would have been irrelevant: a returning salmon was simply a salmon. Since then, the categories of salmon have multiplied. First there is the distinction between wild salmon and escaped farmed salmon. This is significant to many anglers, who see farmed salmon as an uninteresting prey (Nustad, Flikke, and Berg 2010). Farmed salmon can often be easily spotted through signs such as worn fins,[15] but this is not always the case, so scale sampling is needed. Scale sample analysis is the most common method currently used in studies of salmon genetics. Not only does it reveal the individual salmon's biography, but it also gives indications about the salmon's hereditary relations, which are revealed by comparing its genetic profile to a large statistical sample of salmon in a particular river, such as the Vosso.[16]

Within the broad category of "wild," new distinctions are produced, such as the one between cultivated salmon (that is, those that were hatched from broodstock from the gene bank and released as smolts as part of the rescue project) and the rest. The former are easily recognized because their fins are clipped. Fin-clipped salmon are wild in relation to escaped farmed salmon but not quite wild according to the notion of wild salmon as autonomous, able to complete the cycle from egg to egg independent of human intervention. Hence, the cultivated salmon are distinguished from those that have the adipose fin intact (and no visible signs of being raised on a fish farm). The latter could include salmon whose lives have unfolded independently of the project—what project workers, perhaps for lack of a better word, sometimes refer to as "wild-wild" (Dalheim 2012, 99). But it might also include salmon released by the project into the river as fry, or as roe planted in the river. In order to find out which is the case, the fish would have to be killed and cut open, and its brain scrutinized for a colored ring—the result of a marker added to the water surrounding the eggs at the hatchery, precisely in order to leave a sign that the fish is, in fact, not quite wild (Dalheim 2012).[17]

Figure 9 is an attempt to visualize these enactments. Note that this is not a scheme of cognitive categories elicited from informants' verbal distinctions. Instead I have used heterogeneous practices of differentiation as the starting point for making the separations, which are indicated by arrows. The distinctive categories are deduced logically from these practices of separation, and although the various types of salmon that are listed are enacted as separate, they are not always *named* as such, or at least not beyond the practices in which the distinctions are being made.

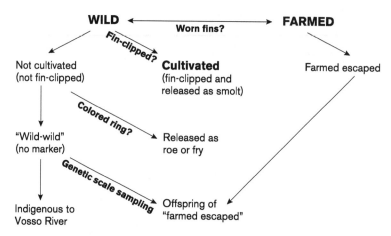

FIGURE 9. Distinctions multiply: An overview of practices to differentiate wild, cultivated, and escaped farmed salmon.

Finally, there is the possibility of taking scale samples. Scale sampling is a common practice across most, if not all, salmon rivers in Norway and is administered by the Norwegian Environment Agency. These various practices of differentiation complement rather than compete with each other and are part of the effort to understand the salmon and its habitat.

In order for the salmon to "speak," a temporary capture is often necessary. This can be done through devices such as the *smoltskrue* (smolt screw), which is "shaped like a funnel and contains a rotating device that traps anything that comes into it into a container" (Dalheim 2012, 87). When the device is brought onshore (usually once a day), the biologists can examine the smolts that have been caught and either let them go back into the river or kill them for further inspection. Another method is a practice called *gjenfangst*, or recapture—in other words, fishing for research purposes. This is done with *kilenot* and *sitjenot*, respectively, fishing devices that have been used in Norway for centuries. A *kilenot* (bend net) is a long fishing net connected to the shore; it leads the salmon into a labyrinth of nets that traps them inside. As with the smolt screw, it can be checked daily. A *sitjenot* requires the presence of a fisherman, who watches from a tiny hut on the steep mountainside by the fjord or river. When a salmon enters the net, the fisherman releases a big stone connected to the entrance, sealing the salmon inside (Barlaup 2008, 136; Dalheim 2012, 86–87). Some returning salmon are allowed back into the river to continue their journey to spawn. Others need to die, because the information they carry can only be revealed from their dead bodies.

Through these techniques, each salmon is made to tell a particular story, revealing a snippet of its origin, life story, or movement. As such stories are gathered,

FIGURE 10. Smolt screw (photo by Marianne Lien)

summarized, and sometimes compared, a particular choreography of salmon movement is produced, describing the Vosso River as a more or less successful salmon river habitat.

So what to make of this? It all depends. River owners may look forward to the lifting of the moratorium on salmon fishing, while conservation biologists are more concerned with biodiversity and preserving indigenous salmon populations. To some extent, one could argue that the various descriptions of salmon in the Vosso happen independently of one another, and the stories of salmon have restricted mobility. Hence, any attempt to combine them in one place is in itself an important move toward a coherent, single rendering of the Vosso salmon. One such effort is the annual meeting at which project participants present their results to the various stakeholders.[18] These meetings are closely modeled on scientific workshops or panel presentations and allow ample time for discussion of the validity and interpretation of novel results and insights relating to the salmon rivers, salmon behavior, interactions between salmon and other species, and so on. Here, the many different stories that the salmon are made to tell are negotiated, contested, and sometimes discarded so that the integrated outcome can be seen as an approximate and provisionally true version of what the Vosso salmon is about.[19]

Another move towards coherence is the regular update of public reports, based on data that are transmitted from various participants, edited by the project leader, and published by the Directorate for Nature Management (Barlaup 2008, 2013). Through such practices, the project becomes a "center of calculation" (Latour 1987), enacting the Vosso salmon as a scientific entity.

All of this indicates that the Vosso salmon rescue project has yielded not only an increase in salmon returns but also a considerable amount of data that feeds statistical and other procedures and contributes to knowing the Vosso River watershed through salmon (and vice versa). Salmon are eclipsed by an ever-expanding vocabulary and by scaling devices, which highlight particular sets of connections and silence others.[20] These are clever and creative tools designed to make the silent salmon speak. And yet, as I follow people tracing salmon in rivers, I cannot help thinking that there are generative forms of salmon that we will never know about. The problem is not that the biologists have an incomplete image of any given river, but rather the opposite: a river becomes overdetermined, generating an overabundance of human-salmon stories of a particular kind. As salmon yield data, they help cultivate an entire river watershed, but what is being rendered visible are fragments of an imagined whole. Put differently, these are the inevitable side effects that are produced through attempts to assemble what John Law (2014) calls a "one-world world."

In the following two sections, I trace the salmon "sideways," along trajectories that disrupt the lines that follow neatly from these cultivation efforts. The first section, about the Arna tributary, traces salmon that take the wrong turn, or "wander off"; I take their lead in exploring some potential hinterlands of salmon becomings. In the second section, about homeless salmon, it is my own imagination that wanders off, departing from the material enactments of human-salmon encounters, as I explore the potential of a multispecies-oriented humanist ethnography.

THE SALMON THAT "WANDERED OFF": THE ARNA CASE

Certainty itself appears partial, information intermittent.
—MARILYN STRATHERN, *PARTIAL CONNECTIONS*

By fall 2011, it was clear that the Vosso salmon rescue project had yielded positive results: salmon returned in great numbers. It was therefore quite a surprise for those engaged in rescuing the Vosso salmon when on 14 October 2011, the county authorities in Hordaland issued a statement: All adipose-fin-clipped returning salmon that had found their way up the Arna River tributary, locally called Storelva, were to be eliminated before the spawning season. The Arna River is a tributary of the Vosso river system located about midway between the ocean and the river mouth.

The statement from the county authorities was a response to reports from Arna River fishermen that as many as 30 percent of the fish they caught had had their adipose fins clipped. Because there are few other cultivation measures in the watershed, they were most likely hatched by the Vosso rescue project and hence "designed" to return to the Vosso, not to the Arna River.

Arna is not known for its natural beauty. It is a rather nondescript suburb of Bergen, 8 kilometers northeast of the city center, and known for industrial plants as well as suburban housing.[21] That the river has any salmon at all is largely thanks to the voluntary activity of the local anglers' association (Arna Sportsfiskarlag),[22] whose members have made various efforts to clean up the river, mitigate toxic spills, cultivate fry, and build a salmon ladder that has a built-in camera. In the 1970s the river was considered "dead" as far as salmon goes. Forty years later, things are looking much better. According to their website,[23] the local anglers landed 299 salmon in 2012, and 2011 was even better. But what kind of salmon are they?

The genetic ancestry of salmon returning to the Arna tributary is a contested topic. Years of industrial development on the river, including several toxic spills and the current construction of a suburban tramline, have nearly wiped out the original native population. But is it only "nearly"? Has the original native population perhaps been *completely* wiped out? Biologists don't know for sure. And does it matter? For some anglers I spoke to, it does matter. They have seen that years of cultivation efforts have brought some results and possibly restored the original Arna salmon. Because there are no genetic records of earlier catches, this is difficult to prove. But photos collected by the Arna Sportsfiskarlag can be interpreted to indicate that the original Arna salmon was smaller than the Vosso salmon and therefore possibly also better adapted to this tributary stream, closer to the coast, with strong currents and quite a few small waterfalls. This was the position of the Hordaland County Governor, who issued a letter to the Arna Sportsfiskarlag explaining the need to eliminate fin-clipped salmon from the river.[24] Others point to the uncertainty over whether this Arna variety ever even existed. In the heat of the controversy, this comment was posted on the Salmon Group's website: "You can hardly become more precautionary than 'emptying a river' *[tømme elva]* for a type of wild salmon in order to protect a variety that no one has verified."[25]

The issue here is *feilvandring,* or "wandering off." The returning Vosso salmon were hatched upstream, more than 50 kilometers east of Arna. They were released to populate the Vosso River, particularly its upstream tributaries, which are excellent for angling. They were never supposed to end up here in the suburbs. That some salmon do not return to their river of origin but "wander off," *feilvandrer,* is well known, and biologists assume that about 5 percent do so. This is thought to make the river system more robust. That cultivated salmon, released as smolts, do not find their ancestral river of origin is also not a great surprise. Biologists speculate that although the general orientation through which salmon navigate to the coast is hereditary, the finely tuned orientation through which they find a specific river may be learned (Jonsson and Jonsson 2002). Most salmon that wander off end up in a neighboring tributary. Because 90 percent of the salmon that returned to the Vosso river system during this period turned out to be from batches released as smolts by the Vosso project and mostly transported by boat through the fjord system towards locations close to the ocean (hence they never made the journey on their own), it is perhaps not so surprising that some took the wrong turn and headed for freshwater at the first opportunity: the Arna river tributary.

But around 30 percent, as was reported, is quite a lot, much more than what might be expected under "natural circumstances" and enough to disrupt any local stocks. Hence the County General's decision to implement measures to eliminate fin-clipped salmon in order to preserve what might be—but is not necessarily—a uniquely adapted indigenous variety.[26]

We could tell the story of the Arna controversy as a conflict of interests between conservation biologists and fishermen, for example, or as a controversy among biologists. But instead I want simply to gesture towards the liveliness of the salmon itself. It doesn't behave according to the script. It wanders off, and it ends up in places where it is not "supposed to be." Who knows what genes it brings along? But indeterminacy is insufficient for county authorities: people need a policy to act on, and the policy is either-or, life or death. In this case, it spelled the death of the Vosso salmon that went astray. Carefully cultivated to help recolonize the Vosso River, some hundred salmon took an unexpected turn and became matters out of place, contaminating rubble, and eventually destined to die: Unlike salmon with the fin intact, fin-clipped salmon caught in the Arna river will not have the option of catch-and-release but will be caught, registered, and then killed, like escaped farmed salmon.

I am not arguing that what happened is wrong, nor would I argue that it is right. I am just paying attention to what is happening to the misfits and the "nearlies" that never made it to the spawning grounds and whose story is rarely told.

...

How can we tell a story that remains sensitive to the indeterminacy of unfolding underwater lives? How can we tell a story that allows incidental rubble to have a voice? How can we add to the stories biologists tell, in a way that does not simply reproduce their precise, but perhaps also excessive, vocabulary?

My impulse is to search for the cracks, the moments of uncertainty when things don't add up. Or to ask questions that others don't ask, which means attending to what is otherwise ignored. In the following story, I rely on the scientific practices of river biologists and fishermen, but I also engage an ethnographic imagination, which means I do not only study practices as they unfold. Perhaps one could say that I follow Edwin Ardener's (1989) intention in his coining the concept "muted groups" by lending an ear to what our knowledge-making practices otherwise ignore.

Let us return to the first distinction, that between farmed and wild salmon, and their troublesome, ever-present shadow: the "escaped farmed salmon."

HOMELESS, OUTLAWS, OR ESCAPEES: DISCURSIVE INTERVENTION

Recall that nearly half a million farmed salmon find their way into Norwegian coastal waterways every year. So what is an "escaped farmed salmon" *(rømt oppdrettslaks)*? For the biologists, it is mostly just that: salmon that was farmed and then escaped. If life inside a pen has left a visible mark on its body, such as worn fins, then the salmon is simply killed and discarded. If there is doubt about its genetic origin, scales can also be sent away for genetic analysis. It might be said that the only "good" escaped farmed salmon is a dead one. Once it is identified as "escaped farmed," there are no further questions, no further distinctions to explore: it is, as it were, a "dead end" in the quest to restore, and to know, Vosso as a salmon river.

But escaped farmed salmon are also more than that. And they aren't necessarily always dead. It took about four years of fieldwork on and off salmon farms before I began to seriously question the term *escaped farmed salmon*. The term is used in official salmon statistics; it is used by salmon farmers and fishermen alike; and although it points to a contested field, the term itself escapes the attention of most people involved, so it escaped my attention too. Then I learned more about salmon trajectories and began to think that an escaped farmed salmon actually never "escaped." Unlike farmed cod, which are notorious for actively breaking out of pens (the Houdini of aquaculture), salmon make few, if any, attempts to break out. Accidental escapes happen because their home at sea, the netting that surrounds them, breaks open. A propeller getting caught in the net, or a similar incident, might ruin the enclosure.[27] Their world simply shatters, and they disperse from that moment. Salmon may have agency, but they hardly "escape." What we do know is that their world is suddenly, dramatically altered.

Within a few hours, they probably feel hungry. It is quite likely that they check out the water surface and wonder where the pellets are. So rather than being escapees, perhaps they are simply hungry, or wandering off, or lost—perhaps they are more accurately referred to as *homeless*.

Some salmon stay close to the pen and continue to feed around the farms.[28] Others move farther away and learn to find their own food. Perhaps we should describe the latter as *refugees,* or *survivors.* The pen is not necessarily a prison; it has been their home so far (their domus), and the waterways that they explore are more than the damaged habitat of their distant native relatives; they are also the place—the only place—where these homeless survivors can get on with the task of living.

But their chances of survival are slim. Within hours, you can see local people fishing from small boats, hoping to fill the freezer in no time at all. The salmon farmers encourage this and offer a small payment for every salmon that is caught. If the salmon make it past the first few days, they may migrate upstream, where other obstacles are waiting: the river anglers and the biologists, with their smolt screws, their bend nets, and other devices designed to catch salmon for research purposes. For the salmon, the result is the same: they die. I wonder whether we could add another label to the vocabulary: *fredløs* (outlaw),[29] protected by no one. They are impure and unfit to take part in the future imagined by those currently defining what the salmon watersheds will look like.[30] In relation to human categories and the structural divide between wild and domestic, they become matter out of place, "dirt" in Mary Douglas's (1966) sense of the term, as aptly expressed by the Namsen rower who was "glad to get that *shit* out of the river."

This intervention is not so much a critique as a reminder that words are political, they are sometimes excessive, and they help make up the world we perceive. Playing with words, as I have just done, invites us to ask other questions and engage different connections.

. . .

By making the "wrong turn," the fin-clipped salmon swimming up the Arna River intervene in the imagined trajectory of the future Vosso River watershed, especially in the cultivation of a possible indigenous stock. The Arna case is an intervention not by words but by the salmon themselves. Renaming the farmed salmon whose net enclosure was broken open as "hungry" or "homeless" or "refugees" or "*fredløs*" is an intervention with words, a refusal to treat these salmon as if their confinement is their most relevant characteristic. It is an intervention related to the seemingly clear-cut dichotomy between wild and farmed in the Namsen River that introduced this chapter. In exposing messiness and "unruly edges," it is an intervention at the level of representation, indicating that "the line between the domesticated and the wild is not as hard and solid as one might think" (Tsing 2012b). It is also an effort to engage in anthropological analysis as more than another ordering device, and instead as a "lateral" theorization or, as Frida Hastrup (2011) puts it, a "matter of showing an excess of perspectives, of cultivating differences, of making ever more things appear, as figures, perspectives and practices bumble into one another in collective life—and thus of pointing to the inexhaustible nature of the social world" (427). Hence, it takes seriously not only that every act of naming, knowing, or classifying is performative (Lien and Law 2011), but that our ethnographic practices are performative too. The implications of this for the distinction between wild and domesticated, between nature and culture/society, are significant and serious. Considerations of what this might mean in relation to the discipline of anthropology and to a more-than-human ethnography have only just begun (see Kirksey and Helmreich 2010; Ingold and Pálsson 2013; Tsing 2013).

Salmon are not either social or natural, but both—or neither. To shift the attention from the precision-seeking vocabulary that emerges in scientific practices to the speculative imagination of a human fieldworker tracing salmon trajectories sideways, across a sampling of ethnographic sites, involves a subtle departure from biology as well as from social anthropology as it is commonly practiced. It also implies a slight departure from the rigid ethnographic commitment to the observation of heterogeneous material practices, as it is practiced in actor-network theory and material semiotics. Are we permitted to describe something we cannot physically see? Are we allowed to speculate beyond the discourse of our human informants? I believe that as we explore the emergence of biosocial becomings, there is no other option.

Tim Ingold (2013) writes that "human becomings continually forge their ways and guide the ways of consociates, in the crucible of their common life. In so doing they weave a kind of tapestry. But like life itself, the tapestry is never complete, never finished" (8). To write ethnographically about salmon and their rivers, and their people, is to write about this tapestry as it unfolds, remembering, of course, that salmon do some weaving too. But more importantly, it means resisting the urge for closure. And this is perhaps where ethnography parts with biology. While the latter seeks closure, sealing the story and the assumption of a "one-world world" (Law 2014), we can work on the assumption that the relational enactments won't all fit together, or that, as John Law puts it, "there isn't a single container universe" and we "need to find ways of puncturing the propensity of the North to self-sealing one-world metaphysics." Following the cracks, the fault lines, and adding a voice that seeks out difference, rather than coherence, is, I think, a way to go.

8

TAILS

So what have we learned about domestication? What is a domesticated salmon? And how can these insights help us navigate a world in which encounters between human and nonhuman are rarely simple and straightforward?

Marine aquaculture offers unique insight into domestication as an ongoing process. We have seen that, far from being settled once and for all, the salmon farming assemblage is indeed a construction site, a "domus in the making." This is not only because this recent transformation makes salmon newcomers to the farm compared to most other husbandry animals. It is also because becoming is an inherent dimension in the ongoing practice of living and being—for humans as well as nonhumans. Domestication is, as archeologists have reminded us, a set of dynamic and mutual multispecies relations.

Rather than seeing domestication as a shift from one steady or stable form of existence to another, I have paid attention to the ways in which every encounter between humans and nonhumans entails a transformative potential. We have seen that such transformations may have occurred long before our historical era and that the outcome is indeed open-ended. "Alien," "sentient," "scalable," and "biomass" are just a few trajectories that may be singled out from within the salmon farming assemblage. One trait does not exclude the others, and the farmed salmon that emerges is indeed "more than one and less than many."[1]

For more than a hundred years, the concept of "domestication" has been a powerful sorting device. Together with its conceptual allies "nature" and "culture," it has created some sort of order in a world of lively encounters. It has also provided humans with guides for action in navigating this world. Wild or not-wild? Domesticated or not-domesticated? These questions help us make sense of messy rivers.

The answer defines the destiny of "homeless" salmon (see chapter 7). In this way, "domestication" performs a distinction, or an intra-agential cut,[2] in worlds that are forever in the making. Can we afford to let it go?

Perhaps we can, but we do not necessarily have to. Because the problem is not "domestication" as such but rather its allies, nature versus culture. As long as domestication draws its meaning from this conceptual contrast, its analytical potential is limited. Suspended between nature and culture, all it can do is endlessly perform a distinction that reifies an outdated order. Although this can be useful for some practical purposes, and for comparison too,[3] it is hardly helpful for ethnographic explorers in multispecies realms. This is because it directs our attention towards ordering rather than curious inquiry.

What we need to do, then, is to imagine domestication without this tired dichotomy—to create a space for domestication practices *beyond* nature and culture. Throughout this book I have approached domestication as relational practices that enact biosocial formations through which humans and nonhumans make space for one another. I have emphasized the transformative potential of such practices rather than their ordering effect. Hence, rather than casting salmon as either wild or thoroughly domesticated, I have suggested that most human-salmon encounters cannot be reduced to such simple contrasts and that these encounters tend to both leave traces and have implications for the future. As long as humans have inhabited the shores and rivers where salmon return to spawn, our histories have been intertwined.

This is not to say that aquaculture is simply a continuation of old subsistence practices or that nothing significant happened. Indeed, the massive expansion of industrial aquaculture and the dramatic decline of self-reproducing salmon rivers during the twentieth and the early twenty-first centuries are dramatic reminders that we are at a crossroads and that caution is called for. Even so, I maintain that searching for a "pure" precontact state of "untouched nature" is neither historically valid nor particularly helpful. For most practical purposes, salmon and humans have "always been together," and our coevolutionary journey is not over yet. The question, then, is not how to restore some ideal order of separation, but how to get on with the task of living in what is already an intertwined, frail, and coconstituted world.

This is how the notion of domestication may turn out to be particularly helpful. Rather than approaching domestication as an outdated modernist narrative of the "fall of nature," we can approach it as a guide to thinking carefully about practices of living well together. But this implies that we abandon the idea of domestication as a single story. Rather than approaching it as an ordering tool of a bygone era, we can see it as a plethora of avenues, roads, and nearly hidden trails to tease out. Again, domestication is not about one trajectory; it is about many.

What I propose, then, is not to completely abandon domestication as a concept that has outlived its purpose but to reappropriate it for the challenges of the present.

The image of the Anthropocene reminds us that there is no nature out there independent of human life. But there is also no human outside of nature. Can domestication prepare us for the "arts of living on a damaged planet"?[4] I believe that it can, but it requires that we abandon the term as a mere ordering device and embrace instead the tension of order and becoming—or of history and process—captured within this notion. In order to do this, ethnography is particularly helpful. Ethnographic stories are generalizations that intervene, or at least have the potential to do so. According to Winthereik and Verran (2012), they have the capacity to "re-present the world in ways that are *generative*" (37; emphasis in original) both for the people and practices they are about and for the authors and audiences that share them.

Domestication is generative too. But what kinds of worlds does it engender? What work does it do? In this book, I have identified two different modes. First, as an ordering tool, domestication creates History. It lays bare the significant transformative moments in a way that makes it easy to distinguish epochs and eras, then and now, or them and us (see chapter 1). This was possible because the history of domestication has been told as *a fundamental shift* in the way humans engaged with animals and plants. Emphasizing control and confinement as key modes of ordering human-animal relations, this evolutionary narrative has even served as an origin story of modern civilization. And through this emphasis on "before" and "after," the domestication narrative laid the foundations for ordering nature and culture too. In this way, domestication serves as a tool for classifying life-forms according to a predetermined scheme.

But this is not the only way domestication can be put to use. As I have showed in this book, domestication can also be approached as a mutual, reversible, and open-ended process. With this approach, domesticating salmon is not so much about transforming the material world (the world does that all the time anyway, with or without our assistance) as it is about participating in the world's transformation of itself and establishing—through such practices of mutual production—the conditions for ongoing growth and development. We may refer to this mode as an approach to domestication as generative process. This is not a complete rejection of the idea that domestication may involve irreversible change, but it adds another layer of inquiry, one that does not take such a change, or indeed a preordered dichotomy of nature and culture, for granted.

In this book, I have taken a snapshot of a moment of great turbulence. If we ever look back at a time when Atlantic salmon underwent dramatic changes, it would be right now, and the hot spot would be the fjords of western and northern Norway. Never before have so many salmon been enrolled in regimes of domestication as right here, right now. Never before have so many salmon been shipped out from a single region to so many distant destinations. Never before has Norwegian salmon been consumed by so many different people around the world. The story of salmon can be told as industrial success or as environmental catastrophe. Both narratives are partly true, yet obviously incomplete. This book is incomplete as well. But instead of stitching together bits and pieces to craft what appears to be a neatly narrated whole, I have told the story of salmon as a story of becomings. I have argued that domesticating practices generate not one single salmon but many. *Sentient, scalable, hungry, alien,* and *biomass* provide contemporary snapshots of salmon in the state of becoming. Together, they offer perspectives, as well as substantial foundations, for reconsidering what domestication might entail, both in general and in terms of aquaculture and salmon in particular.

The challenge, then, is to hold these seemingly contradictory approaches to salmon together and in tension with one another. This is not only another way of arguing that it's complicated but also a way of promoting tension, or friction, as a generative feature of analysis as well as of our day-to-day practice of living. It is a way of arguing that no matter whether you choose, for example, to eat or not to eat farmed salmon, you are neither completely right nor completely wrong. I believe that it is from the disconcerting awareness of such friction, rather than from self-reassuring confirmations that we are "doing the right thing," that responsible interventions and caution may be exercised. And I believe, with Winthereik and Verran (2012), that ethnographic stories can be great tools for making such generative interventions.[5]

. . .

Archeologists have repeatedly insisted that the Neolithic brought fundamental changes and that it paved the way for massive reorganization of social, symbolic, and cultural life. Their insight reminds us that while people are busy securing their livelihoods, their ideas about the world take form simultaneously. Evans-Pritchard's classic account (1964) of the Nuer and their cattle is an illuminating example. "Nature" and "society" are more contemporary examples of ideas that have emerged and that appear real to us, partly as a result of specific agrocentric and, later, agro-industrial forms of domestication. This means not only that our ideas about the world are shaped by the biosocial configurations of the past, but also that current domestication practices—with salmon, for example—will shape our ideas of what an animal is as well as our social interaction with salmon in the future. Consider Tone, the hatchery worker, encouraging the tiny fry to come up and eat (chapter 5). A specific temporal and spatial assemblage makes it possible for Tone to sustain a social relation with her tanks over several weeks. A certain delegation of responsibility and autonomy allows her to exercise her role in a way that attributes sentience to salmon. We may recognize such practices as enactments of care, through which Tone also embodies the role of caretaker and enacts salmon as sentient beings, regardless of what biological scientists have to say on this matter. Salmon is becoming sentient *in practices*—practices that are patchy and noncoherent.[6] Several implications follow.

First, we need to note that domestication practices are embodied, situated, and thoroughly material. Agency is a distributed property of the entire salmon assemblage however we choose to define it. The domus is not a container but an active agent in the practices that shape domesticated relations into being. It follows from this that domestication is patchy as well. Different assemblages generate different temporalities and modes of attachment; we have seen how the hatchery differs from the smolt production site, which differs again from the grow-out site. In all these sites, domestication is done differently; hence the farmed salmon is a situated being too, not radically different from one site to another, but multiple and only partly overlapping.

Second, as we take into account that different assemblages enact or facilitate different modes of being and relating, we need to consider how past and current configurations seriously limit the horizon of our thought. It is notoriously difficult to think "domestication otherwise." As writers, activists, or farmworkers, we can sometimes try to expand horizons in ways that can make a difference. But we also need to be aware that our imagination is shaped by configurations of the past. To make a change requires a conceptual leap and the ability to question what appears at any given time to be obvious or self-evidently true.

To question the obvious is a hallmark of social and cultural anthropology. Twentieth-century anthropologists cultivated this as the art of ethnography and applied it first and foremost in studies of the "human Other." Twenty-first-century ethnographers have cast their nets more widely, and tried—as I have tried in this book—to include nonhumans in the story of the Other. To do this well is difficult. But the effort to tell the world differently seems important for its own sake, in that it insists on ethnography as a vanguard of historical change, challenging the obvious in ever-newer domains. No fieldwork practice can ever provide an immersion in the lives of salmon in the way that it provides immersion in the lives of other people. But we can still cast our nets more widely and admit nonhumans, as Anna Tsing (2013) has suggested, to our notion of the social. This idea is not as radically new as it may seem. Like Evans-Pritchard and his Nuer cattle, for example, I can study animals with, or through, their people. But I can also take his approach a step further: Drawing on insights from contemporary anthropological explorations at the human–nonhuman interface, as well as science studies, I can weave across several other domains, noticing how configurations shift from biology to business and pay attention to frictions as well as gaps, where things rub against one another or do not add up. Hence, when I engage in the hand-feeding of salmon to attract their attention (chapter 3) and when I suggest that farmed salmon in rivers are homeless and never escaped (chapter 7), I invite the reader to take part in an experiment to cultivate our ethnographic imagination so that it may include the salmon, even as they remain elusive and almost always out of sight.

If current domestication practices shape not only our surroundings but also our ideas about the real, then it follows that acting in and on the material has implications beyond the material shaping of our world. Domestication practices work as ontological politics, or perhaps we could say "as ontological politics in disguise" because they are often not recognized as such. By issuing a statement that all fin-clipped salmon were to be eliminated from the Arna River (chapter 7), the Hordaland County Governor made an implicit statement about the Arna River as a future salmon river and as an entity distinct from the Vosso River. In this way, it performed salmon as a particular kind of nature, as well as the limits of Vosso salmon as participants in that future nature. We might say that the statement performed a powerful, although contested, ordering of nature or even a specific domestication of wild salmon.

To attribute all of these implications to the County Governor's intentions would be not only ethnographically incorrect but also unfair. We may assume that, like most of us, the County Governor acted in good faith and according to the knowledge available. Perhaps what he did was a good thing. Domestication practices work as ontological politics more or less regardless of the speaker's intentions. We may control our words but not the broad associations that they carry with them.

. . .

Salmon are eclipsed by a range of stories from the utopian dreams of the blue revolution as the global food solution to dystopian scenarios of rampant industrialization and dying rivers. Together they form a contested and polarized field that is characterized by contradictory messages and mutual distrust. This polarization has defined my role as an ethnographer as well. In Norway, the main focus has been on the impact of domesticated salmon on the wild. When in 2011 the Ministry of Fisheries and Coastal Affairs called for a risk assessment concerning the environmental effects of Norwegian aquaculture, the risk assessment focused on disease (including sea lice), genetic interaction with wild salmon, and local effluence as key environmental risks.[7] In this way, the "environment" was operationalized as Norwegian rivers and seascapes, and the ensuing public debate was equally narrow. Hence, environmental concern was confined to an area that maps onto the geopolitical borders of the Norwegian nation-state. Through this assessment, which works also as a scaling device, Norwegian salmon aquaculture is defined as an environmental problem with the Norwegian public (anglers, nature lovers, river owners) as stakeholders.

Just because salmon paved the way for marine domestication on a global and industrial scale does not mean that they are forever part of that future. Farmed salmon began their journey as "marine scavengers," consuming fish scraps and by-catch that was readily available near fishing ports in coastal Norway. With the exponential expansion of salmon farming in recent years, the demand for feed has increased as well and has implied a gradual shift from the Northern to the Southern Hemisphere. The impact of this expansion on marine resources in the South is as yet poorly understood. The parallel impact of this increased demand on fishing communities as well as local consumers in less affluent regions in the South is hardly even discussed. This lack of concern is mirrored in what I have described as a relative lack of knowledge, and interest in what pellets are made of and where they come from. These examples show how some social-environmental connections are repeatedly mobilized and, similarly, how some connections (such as to Peruvian anchoveta) are systematically silenced. The implication, put bluntly, is a disproportionate concern for anglers and nature lovers in the North at the expense of impoverished fishing communities in the South and a failure to address questions of global social and environmental justice.

Salmon is an icon of wildness, but salmon is also food. It provides livelihoods for people as a commodity today, as it did in the past. In order to address marine resources in the South as well as social justice on a global scale, we need to bracket the image of salmon as either "wild" or "farmed" and approach aquaculture as a question of feed and food. Salmon's insatiable appetite and affluent consumers' insatiable demand for salmon worldwide form a mutually reinforcing nexus, which threatens to disrupt and shift the channeling of global resources. Should farmed Atlantic salmon be gradually replaced by other marine species such as noncarnivorous and more viable fish stocks? A thorough assessment of the implications of an expansive salmon farming industry would require that farmed salmon be approached not only as another domesticated species that exacerbates the global demand for feed resources but also as a protein component of the human diet that could potentially replace, and thus lessen the demands for, other sources of protein. Is salmon better than chicken? Should salmon replace pigs? Does the relative efficiency of salmon (and fish more generally) in metabolizing and utilizing feed, compared with terrestrial domesticated animals, warrant a transition from terrestrial animals to marine animals as sources of animal protein? Does the current need for antibiotics in poultry and pork production (and the challenges with multiresistent bacteria) justify a shift to mass-produced foods that do not require many antibiotics, such as salmon? Or are the current levels of accumulated marine environmental contaminants in salmon too high to make salmon a healthy alternative? And if so, what would it take to make salmon contaminant-free? These questions are beyond the scope of the present account, but they are questions that need to be articulated more forcefully, both inside and outside of the industry.

. . .

Archeologists remind us that most of the transformations associated with human domestication of plants and animals are unintentional. Helen Leach (2007) describes them as unforeseen consequences and estimates that "97 percent of the time since domestication processes began, humans have not understood the mechanisms sufficiently to foresee the consequences for the plants and animals that became their focus, let alone appreciate how they themselves might be changed" (95).

We may have a better understanding of some of these mechanisms than the people living in the early Neolithic, but our current interventions related to domestication of global resources are of a magnitude that far exceeds those of any previous era, as the concept of the Anthropocene also suggests. Put bluntly, our control of the basis for our livelihoods is, at best, only temporary, somewhat contingent, and highly uncertain. To abandon the idea of control in relation to domestication does not mean that anything goes. It means abandoning the image of humans as the sole occupants of the "driver's seat" and to acknowledge that we share that power with other creatures, known and unknown, as well as with the many heterogeneous assemblages that happen to sustain us at the moment. To stretch the metaphor of collective steering, we can imagine a monkey touching the brake, a turtle under the gas pedal, and a mosquito on our left eyelid. Who knows where the staphylococcus reside and what they are about to become? The road is unmarked, windy, and uneven, and we do not know where we are headed. We, and the salmon, would be well served if we were to move slowly and tread carefully.

NOTES

1. INTRODUCTION

Epigraph: Tim Ingold, *Being Alive: Essays on Knowledge and Description* (Oxford: Routledge, 2011), 14.

1. See, for instance, Hard et al. 2008; Menzies 2012.

2. Between 1992 and 2006, aquaculture's shares of global fish oil and fish meal consumption rose from 0.96 million metric tons to 3.06, and from 0.23 million metric tons to 0.78 million metric tons, respectively. This is not entirely bad: salmon utilize feed far more efficiently than the most common terrestrial husbandry animals. Hence, a shift from feeding pigs to feeding salmon could be seen as a better way of using scarce resources. However, because aquaculture does not *replace* terrestrial husbandry but represents an *addition* to it in terms of global demand for wild caught fish, the total pressure is problematic. In 2003 farmed salmon consumed more than half of all the fish oil sourced globally to aquatic animals (FAO 2008, 145).

3. Donna Haraway credits Marilyn Strathern as the one who taught her that "it matters what thoughts we use to think other thoughts," "what worlds we use to think other worlds," and "what relations we use to relate relations." I am indebted to both for generously sharing their insight during a public conversation at the University of California, Santa Cruz, 28 February 2013.

4. Mobilizing domestication as a comparative tool also helps decenter the conventional critique of industrial food production that presupposes a similar dualism between capitalist and noncapitalist or between global and local as key explanatory figures. Such a critique is not invalid, but, as I demonstrate in chapter 5, it is not very helpful in understanding the recent expansion of salmon aquaculture. It is particularly unhelpful in relation to Norwegian farmed salmon production, which performs a kind of capitalism that is (and has been) harnessed, framed, and facilitated in relation to noncapitalist goals defined by the Norwegian state.

5. The term is derived from Viveiros de Castro (2011, 128). While he refers to anthropology as a "permanent decolonialization of thought," I prefer a less totalizing claim, hence the term *ongoing*.

6. The ultimate separation could be seen as the net itself.

7. Zeder (2012) describes domestication as different stages and pathways on a journey in which wild phenotypes eventually move in the direction of "domestic phenotypes." She distinguishes mutualistic relations "in nature" from those between humans and target domesticates, because the latter are "driven by the human ability to spontaneously invent new behaviours and maximize return of a desired resource, and to pass on behaviours through social learning." This, she argues, leads to humans having a dominant role in an increasingly asymmetrical mutualism, which "moves at vastly accelerated pace and carries a much broader impact than any such relationship in nature" (228). From this perspective, salmon aquaculture may be seen as an example of the "end of the journey" through the complete control of the salmon's movement and reproductive cycle, which implies that it no longer migrates to the ocean to grow or returns to the river to spawn. See chapter 3 for a more nuanced discussion.

8. Anna Tsing (2012a) defines scalability as a feature of design that enables something to expand without changing (507). In this perspective, scalability is not a feature of nature—there is nothing about a salmon that is scalable as such—but the outcome of heterogeneous, and sometimes contingent, sets of relations (see also chapter 5).

9. Ian Hodder (1993) points out that its root precedes Latin and is among the most ancient in the Indo-European world. According to Hodder (45), "domesticate" is linked not only to the Latin *domus* and the Greek *domos* but also to the Sanskrit *damas*, Old Slavonic *domu*, Old Irish *doim*, and the Indo-European *dom-* or *dem-*. Hodder further notes that "domestication" is associated with common English words such as *domicile, dominant, dome, domain, dame,* and *tame*.

10. "Domesticate: 1. to convert to domestic uses; tame, 2. to accustom to household life or affairs, 3. to cause to be or feel at home; naturalize, 4. to be domestic." *Webster's Encyclopedic Unabridged Dictionary of the English Language,* new revised edition, 1966.

11. James Scott (2011, 208–13) asks how domestication could lead to population growth, when in fact, at least in the short term, the shift to agriculture led to increased morbidity and mortality (due in part to zoonotic disease). One answer, he suggests, is that grain centers drew in populations from external sources, through forced labor, slavery, or other forms of personal disempowerment. This included measures to deprive such labor forces of alternative livelihoods by undermining or prohibiting food procurement practices (fishing, hunting, gathering) that were not ordered in accordance with regimes of domestication. An example of the latter in relation to salmon is the legal denial of fishing rights to indigenous people such as Ainu (Swanson 2013), Sami (Ween 2012), American Indians/Nez Perce (Ween and Colombi 2013), and Alaskan Aleut (Reedy-Maschner 2011).

12. While Ian Hodder (1990, cited in Russel 2007, 32) has argued that the importance of the domus lies in its power to domesticate people—that is, to control the wild in men by bringing it into the "domesticating female sphere of house and hearth"—Anna Tsing (2012b) blames domestication (especially cereal domestication) for the emergence of private property and of a particular version of stratification within the household that made men paterfamilias and reduced women to instruments for giving birth. Feminist scholars have also

offered a critique of domestication, but that does not mean that they have seriously questioned the basic assumptions that underpin this narrative.

13. To say that domestication is a factish myth is a way of proposing domestication as something that is neither an independent reality that is "discovered" by science nor simply a projection of human beliefs (Latour 2010).

14. As Helen Leach (2003, 249) puts it, "Domestication" has come to mean the process "by which humans transformed wild animals and plants into more useful products through control of their breeding."

15. Taming is often defined as a "relationship between a particular person and a particular animal without long-term effects beyond the lifetime of that animal," while "domestication is a relationship with a population of animals that often leads to morphological and behavioral changes in that population" (Russel 2002, 286), but this does not preclude connections between the two, involving selective pressures that favor individual offspring that have a propensity for human contact—for becoming tame.

16. Russel (2002) cites Bökönyi's definition as a classic example: "The capture and taming by man of animals of a species with particular behavioral characteristics, their removal from their natural breeding conditions for profit" (287). But see Larson and Fuller (2014) for a more nuanced and recent account.

17. The "Neolithic Revolution" started in the Middle East and is assumed to have happened piecemeal and gradually, expanding towards the north of Europe. It reached the southern parts of Scandinavia about 2,400 years BCE. For Childe, domestication was "that revolution whereby man ceased to be purely parasitic and, with the adoption of agriculture and stock-raising, became a creator emancipated from the whims of his environment" (Childe, cited in Leach 2003, 356).

18. In "Retrospect," Childe (1958) described his inspirations as follows: "I at least took over the Marxist terms, actually borrowed from L. H. Morgan, 'savagery,' 'barbarism,' and 'civilization' and applied them to the archaeological ages or stages separated by my two revolutions: Palaeolithic and Mesolithic can be identified with savagery; all Neolithic is barbarian; the Bronze Age coincides with civilization, but only in the Ancient East."

19. Hodder (1990, 271) concludes that "the increased taming of the landscape in the later Neolithic is simply an extension of the older domestication idea and practice, the aim being to extend social control competitively over larger social entities." Hodder (282) linked the household to the domus not only as a built structure but as a conceptual structure.

20. See Viveiros de Castro 2005, 41.

21. One of the most important corrections is that agriculture was not a sudden invention but a long coevolutionary, cumulative process marked by changes on both sides of a relationship in which partner populations became increasingly interdependent (Zeder et al. 2006, 139; Gifford-Gonzalez and Hanotte 2011).

22. Unconscious selective pressures are those that are *not* the result of conscious or intentional breeding for particular traits (as in eugenics and selective animal breeding) but rather result from specific environmental conditions, such as built structures (domus) and changes of diet.

23. While European social anthropology completely branched off from its physical foundations after the Second World War, North American anthropology was still committed to the four-field approach, which involves a much broader concept of the Anthropos.

Although Europeans become (social) anthropologists without learning anything about human evolution or archeology, this is not possible in the United States. However, even in the United States, the four-field approach has not prevented a significant cleavage between the respective fields, which makes these arguments valid even on the American side of the Atlantic.

24. Viveiros de Castro (2005, 51) cites this quote, drawing on Sahlins (1996). Tim Ingold (2011, 7) discusses the same quote from Marx with a slightly different argument about human becoming.

25. According to Scott (2011), "Their movement and subsistence techniques were *designed* to avoid incorporation into the state: their social structure and egalitarian values also served to prevent states from arising among them" (218).

26. David Anderson and his research group in Aberdeen have recently challenged this view. Based on ethnographic and archeological findings in the circumpolar region, they emphasize the hearth and the household as important sites of becoming (Anderson, Wishart, and Vaté 2013). Their approach thus brings attention to mutuality and materials in a way that has inspired my approach as well. I am grateful to David Anderson for conversations on this topic.

27. For ethnographic experiments along similar lines, see Ogden 2011.

28. But see Anneberg 2013 for a more nuanced account.

29. What does it mean to say that something is "alive"? And how does it matter? Anna Tsing suggests that "living things include futures in what they do in the present" (2013, 28; see also Kohn 2013) and proposes that the notion of sociality can be extended to the living (but not, for example, to rocks). I find this distinction useful for those instances when being alive makes a difference.

30. For an example of how such sites can be approached as ethnography, see Asdal 2014.

31. My engagement with Tasmanian salmon aquaculture involved a focus on transnational biomigration and global connections (Lien 2005, 2007a, b).

32. In Tasmania, I had heard global experts refer to Norwegian salmon aquaculture as "state of the art" (Lien 2007b), and I had often been mistakenly seen as "an expert" simply because I was Norwegian. My ignorance is, nevertheless, rather typical: aquaculture is not something that most Norwegians know much about.

33. "Newcomers to the Farm: Atlantic Salmon between the Wild and the Industrial" involved John Law and me, who did fieldwork on salmon aquaculture; Gro B. Ween, who did fieldwork on the Tana River; Kristin Asdal, who did an analysis of cod farming based on written documents; and three master students, Line Dalheim, Merethe Ødegård, and Anita Nordeide, who did fieldwork on salmon in the rivers Vosso, Alta, and Namsen respectively. In addition, we collaborated with biologists Børge Damsgård and Sunil Kadri, and veterinarian Cecilie M. Mejdell.

34. I also draw on work that we have published together: Law and Lien 2013, 2014; Lien and Law 2011.

35. Let me draw on another example from the field of marketing. In a marketing department where I did fieldwork in the early 1990s, popular jargon often described "the market" as a virtual battlefield, with the product managers as "warriors" fighting for their products' long life on supermarket shelves. An analysis of such jargon would support common economic models of markets as anonymous and highly competitive arenas, and the local jargon

could be seen as an indication that such models had been internalized. However, as I tailed my interlocutors in and out of meetings, elevators, and boardrooms, I soon realized that the description of markets as battlefields hardly captured the professional approach of the product managers. Rather than acting as "warriors," they could just as easily take on the role of practical philosophers, allowing uncertainty, doubt, and social affiliations and friendships to emerge more often than a conventional economic model of markets as battlefields would indicate. These subtle nuances were operative "under the surface" of a rather aggressive jargon, and it was only through keen attention to practices and off-the-record chat that they became available to me ethnographically (Lien 1997).

36. As Bubandt and Otto note in their introduction to *Experiments in Holism* (2010, 2), anthropological holism has often been a postulate about, rather than a search for, wholes, but this does not mean that this slightly awkward term is no longer relevant. Hence, they want to explore the possibility that "anthropology can be holistic without being totalizing, [and] that there can be holisms without wholes" (10).

37. For the role of amazement (*forbløffelse*) as a key analytical tool, see K. Hastrup (1992); for a more recent advocacy for anthropology as driven by the spirit of adventure, see Howell (2011).

38. Marisol de la Cadena (2014) proposes the term "ontological opening," which suggests not a radical theoretical shift but a foundational opening towards matters taken for granted, matters to do with assumptions of the real.

2. TRACKING SALMON

1. Aina Landsverk Hagen, January 2013.

2. Wild Atlantic salmon typically spend some two to five years in the ocean before they return as adults to hatch in—or near—their river of origin. In the meantime, they have been putting on weight, from 2 or 3 to nearly 20 kilos, and it is at this stage that salmon are valuable as a food source, sustaining communities along the Atlantic rim. A similar process applies to Pacific salmon (Lien 2012).

3. Spot prices manifest in different ways. One important reference is Fish Pool ASA, a stock exchange for trading financial salmon contracts, forwards, futures, and options. Fish Pool serves as a mechanism for distributing the risk involved in growing salmon for a global market with fluctuating prices. Its website (http://fishpool.eu/spot/) provides regular updates of global spot market prices for farmed salmon as well as forward prices.

4. All company names are fictitious, as are the specific names of localities. With a few exceptions, most place names are real.

5. Sjølaks's annual production was nearly thirty million tons of salmon. This is more than twice the total annual output of Tasmanian salmon aquaculture, but in a local context, it accounted for only about 3 percent of the total annual production of farmed salmon in Norway in this period.

6. Added together, our on-site fieldwork working directly with salmon was only a few months. In addition, because our research took place over more than four years, there were also many moments of fieldworking in between: on the phone with informants, on the web, by e-mail, and in meetings, where I sometimes presented preliminary findings. This prolonged on-and-off fieldwork strategy added insight into the dynamic nature of the company

and the industry, and we developed close affiliations with a few people over time. Our fieldwork in Norway would have been far more difficult to carry out had I not already known quite a lot about salmon farming from previous work in Tasmania. In this sense, both John and I profited from my familiarity with similar sites elsewhere.

7. This asymmetry became clear to us when on one occasion, a Scottish team of vaccinators was hired for a couple of weeks. Having spent a lot of time in Scotland, John became the talker, while I was left out of the conversation, due to my inability to grasp the Scottish dialect over the constant noise from the vaccination machines.

8. Anadromous fish include salmon, bass, and sturgeon (NOAA, Northeast Fisheries Science Center, 2011).

9. In Norwegian: "Ganga skal Guds gåva til fjells som til fjøre, um ganga ho vil" (Treimo 2007, 30).

10. Loke is a central figure in what is known as the older Edda, or the "poetic" Edda, written down in Iceland during the thirteenth century and drawing on myths that are much older. *Jotne* are mythological creatures of Norse cosmology who inhabit the mountain region *Jotunheimen* in south-central Norway.

11. Erna Osland was married to one of the early salmon cultivators. Her book about pioneers in Norwegian aquaculture is a rich personal testimony about the development during the early years.

12. Why this shift from trout to salmon? Part of the reason is the value. Salmon not only was seen as the "king of fish" but it also commanded a high price at the time and was a much-wanted export commodity, possibly partly as a result of the success of the North American canning industry in the early twentieth century, which distributed canned salmon across much of continental Europe and the United Kingdom (but not Norway).

13. Incidentally, almost exactly the same questions were asked in relation to the discovery of oil in the North Sea, which happened around the same time, and the emphasis on fairly tight regulation as well as on public investment in research was a similar strategy.

14. The question of regulatory governance caused considerable controversy for the Lysø committee. Was aquaculture an extension of fisheries, and hence to be handled by the Department of Fisheries, or was it another kind of agriculture and the responsibility of the Department of Agriculture? A minority voted for the latter, arguing that most challenges in aquaculture would have to do with disease, reproduction, feeding, and nutrition, all of which were within the purview of the livestock sciences. However, they were outnumbered by the majority, who emphasized instead the connection between aquatic animals and fisheries (NENT 1993, 17).

15. The committee proposed a maximum annual output of 50 tons per site, but this was never implemented (NENT 1993, 21).

16. With a production of around 150,000 tons a year in 2010, Scotland is the third-largest producer of farmed Atlantic salmon (after Norway and Chile). Slow growth can be a good thing, and in recent years Scottish farmers have spearheaded organic aquaculture. According to the Scottish Salmon Producers' Organization, in March 2014 70 percent of all Scottish salmon had the British RSPCA Freedom Food accreditation (http://scottishsalmon.co.uk/freedom-food).

17. The Lysø committee predicted a production of between 8,000 and 15,000 tons of salmonids in 1985. However, by the end of the year, the actual numbers were nearly double:

29,000 metric tons of salmon and 5,000 tons of rainbow trout. During the next three decades, the total production in Norway grew to more than a million metric tons (see also chapter 5).

18. The Norwegian salmon farming industry often takes some credit for having established salmon aquaculture in Chile; this is undoubtedly partly true, but it is not the whole story. Anthropologist Heather Swanson (2013) notes how the Japanese also played a key role in paving the way for Chilean aquaculture by establishing Chilean hatcheries to ensure fresh salmon for home consumption. This became necessary after new ocean-fishing regulations complicated Japanese access to salmon in the North Pacific. Once the Japanese hatcheries in Chile were in place, this was followed by Norwegian capital investments in the 1980s and a transformation from sea-rearing hatchery practices to enclosed salmon aquaculture.

19. The research station later became Akvaforsk.

20. There are two main suppliers of fertilized eggs in Norway: Aquagen and Salmobreed. On its website, Aquagen (2014) describes its products as follows:

> Today's Atlantic salmon products are based on the collection of varied and representative genetic material from 41 salmon rivers in the period from 1971–1974. Through the years of continuing development work a genetic material has been created which is suited to today's standards for producers and consumers.
>
> In the breeding nucleus we have 600 families which give a broad basis for the selection of breeding fish for production of quality eggs for the market. Family-based breeding work and selection through 10 generations is the foundation for being able to offer a fish which grows fast, is vital, robust, and has a fine colour and shape. This is achieved through the systematic measurement of more than 20 traits of about 100,000 fish in each generation. Each egg batch is fully traceable and documented.

The most important traits are growth, flesh quality, and resistance to disease.

21. Between 1979 and 1989, the FCR (feed conversion rate), which indicates the amount of feed required to produce an equal amount of fish, fell from 2.2 to 1.5 (NENT 1993, 40). Today the FCR is usually even lower, and closer to 1 (see chapter 4; see also Lien 2007).

22. According to one story, Skjervold was already visiting the Australian mainland, attending a cattle conference.

23. Only later did they discover that this was not the case: the estuaries were in fact rather shallow and had limited current, and the salmon farms were eventually placed further out.

24. Lien 2009.

25. During the years 2012–2014, the HAVBRUK program spent about 100,000 NOK annually on aquaculture research (Research Council of Norway, 2011, 14).

26. Paraphrased from fieldnotes.

27. Although most returning salmon find their way to the river where they hatched, an estimated 5 percent return to a different river (for details, see chapter 7).

28. Actually, the joke was from Douglas Adams's *Hitchhiker's Guide to the Galaxy* and is not about God but about Slartibartfast, a designer of coastlines, who won an award for the

fjords on the coast of Norway. For Earth Mk. II, he is assigned Africa and is unhappy; he wants to make more fjords because "they give a lovely baroque feel to a continent" (*Hitchhiker's Guide to the Galaxy: Original Radio Scripts,* Google e-book, http://books.google.com/books?id=6LgGIvmSceoC&dq=%22I%27ve+been+doing+fjords+all+my+life%22&source=gbs_navlinks_s).

29. Similar stories about garlic as a remedy for sea lice were often told among salmon farmers. It is hard to say where the idea came from, but in rose gardens in Norway garlic is a common remedy to combat lice. While garlic is a common food product today, in the 1960s and 1970s it was still exotic and not commonly used. Hence it was known for its medicinal properties and for the foul breath that would betray its eater even a day after it was consumed.

30. This and all other dialogues are translated from Norwegian.

31. The appreciation of what is called "peace and quiet" (*ro og fred*) is not unique to aquaculture. Marianne Gullestad (1992), in her analysis of everyday life in Norway in the 1980s, devoted a whole chapters to a discussion of the cultural meaning and appreciation of peace and quiet and concluded that it is a valued commonly used to justify actions without itself requiring any further justification.

32. Norway has had a relatively high proportion of women in paid employment for several decades, compared with other OECD countries. In 2011, 73 percent of Norwegian women aged 15–64 were employed, about 16 percentage points above the OECD average (Organization for Economic Co-operation and Development 2012).

33. See, for example, Kjærnes, Harvey, and Warde 2007.

34. Most other salmon-producing countries are seen as more liberal than Norway when it comes to state surveillance and regulation. In Norway, this is attributed to what Norwegian industrial actors see as a more neoliberal governance of the private industrial sector elsewhere. Examples include genetically modified salmon, which are permitted in North America but not in Norway; antitrust regulations in Australia that promote competition rather than collaboration; and the industry collapse in Chile in 2007, when the virus Infectious Salmon Anemia (ISA) wiped out more than half of Chile's salmon stocks. The latter was attributed mainly to a lack of regulatory measures during a period of rapid growth. As a report commissioned by the Global Aquacultural Alliance (and cosponsored by the World Bank and Chilean authorities) concludes: "This impressive technical and commercial success was not accompanied by matching research, monitoring and regulation to guard against foreseeable biological risks" (Alvial et al. 2012, 72).

35. During the past several decades, Norway (like the other Nordic states) has been known for relatively high equality, measured as income disparity between the rich and poor. More recently, however, the income gap between rich and poor has increased according to OECD figures (see, for example, Organization for Economic Co-operation and Development, October 2008, "Are We Growing Unequal?" www.oecd.org/social/soc/41494435.pdf).

3. BECOMING HUNGRY

1. The amount of feed fed to the fish is measured in kilos as it is transmitted from the silos through the feed pipes. Estimates of the average number of fish in each pen and their

average individual weight are entered regularly, as biomass. The amount of feed fed the fish each day will adjust the average biomass automatically, on the computer (assuming that the fish actually consume the feed), and this in turn gives the operation manager an indication of how much feed they need to distribute the following day, or week. For more on feeding and biomass, see chapter 4.

2. The fins of farmed salmon tend to get worn down; see chapter 7.

3. *Leppefisk* is the generic Norwegian term for six different types of wrasse that are used as cleaner fish, including *berggylt* (*Labrus bergylta*, Eng: Ballan wrasse), *bergnebb* (*Ctenolabrus rupestris*, Eng: goldsinny wrasse), and *grønngylt* (*Symphodus melops*, Eng: corkwing wrasse). There has been a rapid increase in the use of leppefisk as cleaner fish on salmon farms in Norway since 2009. In 2010, the landed catch of wrasse in Norway was 440 metric tons. This was more than twice the 2009 catch and nearly ten times as much as the landed catch of wrasse in 2008 (Espeland 2010, 4). Concern emerged about the effect of the increased demand on the local wrasse populations, and trials to raise wrasse were instigated.

4. Young smolts are particularly vulnerable to sea lice, and a high number of lice will kill the fish.

5. The local price for *leppefisk* was around 6 kroner each.

6. The moratorium applies to fishing for salmon in the fjords and near the sea. River angling is allowed only during the summer season, and this has given rise to some controversy. See also chapter 7.

7. See, for example, Latour 1987.

8. The medium affords movement and perception, according to Ingold (2011, 22–23). I would go much further and add that for mammals as well as for fish, it constitutes the core of breathing, it conveys the oxygen that is continuously transported to our cells and without which life is unimaginable. Oxygen is not just transported by the medium air, but it constitutes the very essence of air on earth, just as it constitutes water. Hence rather than simply being mediums in which we move, see, and operate, air and water are constitutive features of the living being, of humans and fish respectively.

9. Affordances are "particular qualities of things through which an object lends itself to specific relational possibilities according to whether it is hard or soft, sharp or blunt, liquid or solid, pliable, malleable, or rigid" (Harvey and Knox 2014, 7; Ingold 2011).

10. Pollock sometimes enter the pens while they are young and put on weight until they are too large to exit through the netting. But even if they remain outside, they still graze near the salmon farms, and fishermen worry that this leads to poor quality and reduced demand (Otterå and Skilbrei 2012).

11. Knowing salmon is a recurrent theme throughout the book. See especially chapters 4 and 6.

12. I refer here to Despret's (2013) critique of the concept of presence: "This abstract term—most of the time under the guise of the 'presence of the observer'—while referring to the body, actually conceals it. It conceals what the actual and concrete 'presence' is for the animals: the space the so-called observer's body occupies, the body which moves, which walks, bears and diffuses smells, makes noise, follows, and does everything a body may do—including what we don't know our body may do since we are so unaware about what it is capable of, but which animals may nevertheless perceive" (52).

13. Participant observation, the hallmark of anthropological fieldwork, has always embraced this idea, but usually with a focus on humans, and including animals only as their human relations may be observed by the "disembodied observer," but see Remme 2014.

14. See also Grasseni (2004) and Ingold (2000) for related descriptions of enskilment in human animal relations.

15. For example, most forms of handling fish are seen to have negative impact on their appetite. If they have recently been treated against sea lice, counted, or moved from one cage to another, it is taken into account and explains a change in feeding. Hence, what they look for are changes in appetite that *are not otherwise accounted for.*

16. All permanent staff members are entitled to five weeks of vacation each year. Four of these are typically taken in the summer.

17. Higher temperatures cause a decrease in oxygen saturation in the seawater.

18. Veterinary services are partly private and partly organized under the Norwegian Food Safety Authority, which is a national administrative body with local offices all over the country. The role of the Food Safety Authority is to propose and manage legal regulations and risk-based supervision and to disseminate and implement systems for quality control in the food market and in food production.

19. The Norwegian word *sture* means something like "sulk" or "mope," as when a child is unhappy and inactive, or when a plant or animal doesn't thrive. The word is mostly used about people and has an emotional dimension, usually indicating sadness or a light depression.

20. IPN, or *infeksiøs pankreasnekrose* (infectious pancreatic necrosis), is one of the more common viral diseases in salmon aquaculture. Like pancreatic disorder (PD), it is prevented by vaccines, but the vaccines are not completely effective.

21. This form has three categories: *faste* (firm), *bevegelige* (mobile), and *hoa* (mobile female). *Hoa* is vernacular dialect meaning "she" in plural, or "shes." Sea lice are classified according to sex and stage of life cycle. The main concern is the number of sexually mature females, or "mobile females" *(bevegelige hoa,* or just *hoa).* Other categories are males, which are simply classified as "mobile" (*bevegelige*), and "firm" *(faste),* which are sea lice that are not yet sexually mature. Hardanger Fiskehelse Nettverk was a three-year research effort aimed at determining the interactions of salmon lice between farmed and wild salmonids in the Hardangerfjord system of Norway. It was organized as a collaboration between the Institute for Marine Research, local branches of the Food Safety Authority, and salmon farming companies in the region.

22. Zoning is a public requirement that fairly large areas be left fallow for at least six months on a regular basis. Zoning involves a system of rotation across companies, which can no longer distribute smolts to their own licensed localities but need to coordinate their activities with each other.

23. Sea-lice treatment is done by "bathing" with various types of medication. It is costly and labor-intensive, and depending on the kind of medication, it represents a risk either to the environment or to the health and safety of the workers. In addition, there are signs of resistance to the current medication. In spite of this, there is a general agreement on the need to do it. It is seen as a precautionary measure for the salmon stocks and those passing by, rather than for the ones that are penned up, for which the thresholds are unnecessarily low. (For more details on delousing, see Law and Singleton 2012). According to my

fieldnotes from an interview with a veterinarian in January 2010, the four most important forms of medication in 2010 were AlphaMax, Salmosan, Slice, and chitin inhibitors.

The threshold for treatment varies during the year. Between 31 January and 31 August 2010, the threshold was 0.5 mobile females and 3 mobile males per average salmon. Between 1 September and 31 December, the threshold was 1 mobile female and 5 mobile males. During the two weeks in summer when smolts are passing by, thresholds are 0.2 mobile, or 3 lice in total. If these thresholds are exceeded, delousing is mandatory. In addition, there is a coordinated delousing across all localities in the region scheduled at the end of March.

24. This is very different from the situation in Chile, where the use of antibiotics has skyrocketed in the late 2000s. According to Cabello et al. (2013, 3): "While Norway, the United Kingdom and Canada used approximately 0.0008 kg, 0.0117 kg, and 0.175 kg, respectively, of antimicrobials for each metric ton of salmon produced in 2007, Chile used at least 1.4 kg per metric ton." In 2009, there was a severe outbreak of ISA (Infectious Salmon Anemia) on salmon farms in Chile, with huge impacts on the entire salmon industry. Salmon farmers in Hardanger commented that this was a crisis waiting to happen, due to lack of hygienic measures. It was attributed to what was seen as an irresponsible regime of production that prioritized short-term gain at the expense of ensuring fish health in the long run. A lack of state governance and a weak system of food safety control were also mentioned as contributing factors. During the disease outbreak, the *New York Times* reported that Chile's Economy Ministry revealed that "Chile used almost 718,000 pounds of antibiotics in 2008 and more than 850,000 pounds in 2007." Compared with information from the Norwegian Institute of Public Health, this "was about 346 times the amount of antibiotics Norway used in 2008 (2,075 pounds), and almost 600 times the amount Norway used in 2007" (Barrionuevo 2009). According to the Norwegian Food Safety Authority (2014), the total use of antibiotics in Norway in 2013 was 972 kilos (the total production of fish was more than a million metric tons).

4. BECOMING BIOMASS

1. Feed costs increase as the salmon grow but rarely account for more than around 50 percent of the total cost of production. The other expenses (such as the cost of vaccination, veterinary care, delousing, infrastructure investments, and labor) are more or less constant, regardless of the size of each fish.

2. Kontali Analyse 2007.

3. For details on the collapse of the Chilean salmon farming industry, see chapter 3, note 24.

4. See, for example, Paxson (2013). For such products, profits can be made by efforts to enhance or control specific aspects of the product related to flavor, production practices, origin, or even packaging.

5. This implies that there is little incentive for product differentiation among Norwegian producers. A salmon is a salmon. This is particularly true for Norwegian salmon, which is marketed globally on only the basis of country of origin. In Scotland, where there is a much greater demand for organic salmon, a wider selection of salmon exists in supermarkets.

6. The Industry Standards for Fish (1999) applies the following definitions to farmed salmon. Superior: "A first class product which makes it suitable for all purposes. The

product is without overall faults, damages, or defects, and provides a positive overall impression." (A range of specific criteria follow.) Ordinary: "A product with limited external or internal faults, damages, or defects. The product is without faults, damages or defects that make future use difficult." (Again, a range of specific criteria follow.) The category of Production is for those that do not fit the criteria of Ordinary or Superior. These are sold with heads off.

7. The lack of predictability makes it difficult for the buyers and sellers to plan their investments and operational activities in a longer time perspective. The industry needs an instrument for risk management that can offer both better predictability for the bottom line and the flexibility needed to trade a biological product. Fish Pool ASA (see chapter 2, footnote 3) serves as a mechanism for distributing the risk involved in growing salmon for a global market with fluctuating prices.

8. In Norway, employee participation is inscribed in the legal system as well as in the shared understanding of the relation between employer and employee, which is often described as one of relative equality. It is rooted in a key agreement between the main employers' association and the main labor union, dating back as early as 1907, but also in several attempts during the 1970s to strengthen workplace democracy and participation (Brøgger 2010, 480–81). This participatory and egalitarian ethos is a key topic as well as a gatekeeping concept of regional studies in Norway (Lien 1997; Barnes 1954).

9. There is a great deal of respect for the owner but also a sense of informality. Always called by his first name, he is often jokingly referred to as *høvdingen,* or "chief."

10. Tore didn't know it yet, and neither did I, but it would soon get worse: in a few months he will spend New Year's day and much of his Christmas here trying to protect young fish from the devastating effects of an exceptionally cold winter. See also the section "When Death Is Unexpected: Notes from the Emergency Team" in chapter 6.

11. One time they even tried raising turbot but gave up. When it was built, this was one of the first and most modern smolt-production sites in the region, and Tore tells stories of how they used to mix fish scraps in a cement mixer and use garlic as a remedy against lice. Similar stories are told elsewhere by other people who remember the 1970s; see, for example, Rebecca in chapter 2. As is the case for Rebecca, Tore's familiarity with the site and the way it works precedes the current ownership. While smolt-production sites and grow-out sites occasionally change hands and become part of larger economic units, there is often significant continuity "on the ground" through the local workforce, whose loyalty and sense of belonging is linked to place and community rather than to the shifting configurations of legal ownership. This continuity also ensures skill development and work experience and appears to give the industry a competitive edge. Hence, while the aquaculture industry is a global enterprise, it has some qualities that are clearly nonscalable; see Tsing (2012).

12. The black roof facilitates control of light; for details, see chapter 5.

13. This is, of course, not entirely true: grow-out sites have their challenges, and the potential economic loss associated with a disease incident, for example, can be enormous. Hence, the weight of responsibility attached to numbers, and thus also to operation managers, is significant. But for staff members that were not directly accountable for these numbers, it seems that their production system was somehow less at risk. Staff at Vidarøy rarely had their sleep disrupted by alarms.

14. Aluminum levels fluctuate with the amount of rain.

15. Damage to the aerators also causes the alarm to go off. In addition to these parameters, the alarm is set up to go off in case of generator failure, a pressure drop in the pipe leading into the water-circulation system, water pump failure, or a critically low water level in the dam upriver.

16. The oxygen increases automatically because the tanks are connected to an oxygen tank outside, and the oxygen delivered to the fish tank is turned up when oxygen saturation in the water decreases below 85 percent. The oxygen levels in the outside tank are, in turn, electronically reported to the supplier, a firm in eastern Norway. In this way, the oxygen supplier can monitor the oxygen consumption at Frøystad and schedule its deliveries on the west coast so that the tank is replaced before the oxygen runs out. Thus we can imagine that the turning off of a light switch is linked to the scheduling of oxygen tank deliveries through a set of translations that involve the oxygen metabolism of a batch of startled smolt and weave together smolt responsiveness (and possibly fear) with trailer delivery schedules and industrial work shifts.

17. A scale of 1 through 5 is used to indicate whether it is sunny, partly sunny, overcast, foggy, etc. A similar 1–5 scale indicates how windy it is.

18. In standard written Norwegian (*bokmål*), "the other side" is *den andre siden*.

19. See Latour (1987). The term "center of calculation" is associated with the notion of "cycles of accumulation," denoting the accumulation of knowledge situated within particular locations connected within a network. According to Latour, cycles of accumulation are constitutive of centers of calculation.

20. Or rather, perhaps a way in which norms and expectations can be communicated without being made explicit. Ambiguity, storytelling, and rumor make it possible to convey the nuances. The teenagers who came to help out for a few days but did not pay proper attention were referred to as *gutane*, or "the boys." Some women whose actions are not endorsed become *kvinnfolka*, or "the womenfolk," which can have a slightly derogatory connotation in Norwegian.

21. The paperwork that is prepared when fish is transferred from one operational unit to another includes detailed information on, among other things, the number of fish, their origin (firm, site of production), time of fertilization, age of broodstock (male and female), temperature at incubation, degree-days at hatching and first feeding, average weight at point of delivery, vaccination, statements about health control, licenses, the delivery firm's permissions to produce smolt or eggs, a stamp from the local veterinarian confirming this, and a signature by the operation manager responsible for delivery.

22. This calculation is not always simple: recall Fredrik and his struggles at Vidarøy to calibrate the automatic feeders so that the numbers on his computer would accurately measure the amount of feed actually pumped through the system (see chapter 3).

23. While "economic FCR" includes the count of morts (*daufisk*) in the given period, the "biological FCR" is calculated without taking morts into account and thus reflects more accurately the metabolic efficiency of current stock of fish in the pen. A sudden increase in mortality will, for example, cause a rapid rise in economic FCR (feed has been "wasted" on fish no longer contributing to biomass) but may not have the same impact on the biological FCR.

24. In Tasmania, smolt-quota performed a similar function, enabling authorities to place an upper limit on the total production.

25. In Norwegian: "Den til enhver tid stående biomasse av levende fisk, målt i kilo eller tonn. Biomasse er oppgitt i levende vekt" (Directorate of Fisheries [*Fiskeridirecktoratet*] 2009).

26. Altinn is Norway's unified public electronic reporting system. Originally established primarily for businesses to file required reports, it is now widely used by private individuals as well, especially for filing tax returns.

27. For details, see Lien 2007, 177.

28. They are also at the stage when they need to be vaccinated, and in order to minimize handling, these things are often done at the same time. An automatic vaccinator not only vaccinates, but weighs and counts as well, and if you set it up right, you will know exactly how many have come out at the other end as well as their average weight. For details, see Law and Lien 2013.

29. The "winter regime" refers to the process of smoltification, and in this case a technique for producing smolt twice a year rather than just in spring. This means that smolts scheduled for delivery in September are placed under limited light, to speed up the smoltification process. Chapter 5 has more on temporality and smolt.

30. The e-mail transcript of these sentences: "Men det er som alltid feilkilder; Storfisk ha en tendens til å svømme djupere en de små som har tendens til å svømme langs notkanten. Plassere du måleren djupt får du en annen snittvekt osv."

31. The e-mail transcript of this last comment: "Vi pleier å treffer ganske godt, de fleste regner med at det er mer biomasse i en merd enn det er i virkeligheten og det resulterer i en skuffende fôrfaktor når fisken slaktes."

32. I happily accept his proposed distribution of labor: lifting fish that weigh around 5 kg out of the water with a dip net is a heavy task, especially since the water surface is below our feet and the dip net's handle is several meters long. Karl is broad-shouldered and his upper arms are stronger than mine.

33. No wonder Karl sneers at the office. Last night, after we had dinner at his house, he handed me his tax form, which needs to go the head office. I had to deliver it for him, because when they depart for Vidarøy in the morning, the office is not yet open.

34. I notice how the word *paper* keeps popping up in conversations and in my own questions, even though paper has very little do with this.

35. Through a special function, the Fishtalk program also produces the form required in the Altinn system. But internal quality control is still required, and at Sjølaks, this is Finn's responsibility.

36. If a company exceeds MTB (maximum allowable biomass), it could be punished with what Finn describes as "a terribly high fine" (*en forferdelig stor bot*). At the time of these interviews, hearings were being held among local stakeholders on a new regulatory amendment for Hardanger (*Hardangerfjordforskriften*). It proposed that the total MTB for the entire region be readjusted from the current 70,000 metric tons to 50,000 metric tons. The difference would impact most producers and effectively cap further growth in the Hardanger region. While the aquaculture industry, currently geared towards further expansion, was critical of these measures, environmental organizations advocated an even lower maximum threshold (see also chapter 2, "The Aqkva Conference: A Transient Patch").

37. Europe is the main market for Norwegian salmon, but Asian countries are important as well. The amount sold to the United States is negligible in comparison, eclipsed even

by domestic (Norwegian) consumption. Farmed salmon in the United States is primarily Chilean or North American and in competition with wild or sea-ranched Alaskan salmon (Kontali Analyse A/S 2007). For details on the branding of Alaskan salmon, see Hébert 2010.

38. Bids are accepted on the phone, and oral agreements are binding, but they are confirmed by a written statement (e-mail or fax) at a later stage.

39. Larger companies often establish an in-house exporting office, and thus they bypass independent exporters.

5. BECOMING SCALABLE

1. Value adding has been the common strategy during the last decades in food manufacture industries that cater to the so-called saturated markets of Europe and the United States. In these relatively affluent markets, although there is an obvious limit to how much people can eat, there is still a potential for economic growth in that consumers can be willing to pay more for the same amount, provided that it the product has some perceived "added value," symbolic or otherwise (Lien 1997).

2. In 1985, Norwegian salmon was sold on the global market for the equivalent of more than 80 kr/kg. Twenty years later the prices were less than 30 kr/kg (Asche, Roll, and Tveterås 2007, 53, figure 12).

3. 1,165,954 metric tons, to be precise (Statistics Norway, 2014).

4. I thank Bernt Aarseth for his advice in assembling these figures.

5. Norway's leading role as a fish exporter has traditionally relied on capture fisheries. However, while capture fisheries are still important, aquaculture has grown much faster and is now the largest source of Norway's fish exports (FAO 2008, 48, part 1, figure 8).

6. This adds up to 59,720 truckloads (each with 18 metric tons of fresh whole salmon) a year. The number is an approximation, assuming that all salmon is for export and that all is sold fresh, which is not the case. However, it is still useful for visualizing the magnitude of the production (Are Kvistad, FHL, personal communication, March 20, 2012).

7. The idea, as it is explained by the hatchery workers, is that in the river, very few eggs will become smolts. The hatchery provides an extra protective layer, which gives the weaker ones a chance to survive a little longer than they would have in the river. But in the long run, they are not likely to make it through to adulthood. Culling those that are not perfect is a way of slightly diminishing the selective bias created by the hatchery conditions and of making the batches more uniform and healthy.

8. The surface is crucial. Many versions exist in commercial hatcheries, and one of them is AstroTurf. (AstroTurf is the brand name used generically for the short-pile synthetic turf invented in the 1960s for use on playing fields. The kind used at the hatchery at Idunvik has thick plastic fibers; it is also used in Norway as a floor covering on balconies and other semi-outdoor places, where it adds softness to the ground in a way that resembles a lawn.) AstroTurf is used to hold the fry until they start feeding. Olav thinks it is a great invention. He cannot recall when he started using it, but he remembers discovering that the alevins on AstroTurf grew much more quickly. Before what he calls the "AstroTurf era," they would be about 0.14 or 0.15 grams. With AstroTurf, they sometimes get as big as 2.5 grams in the same

number of degree-days. Olav thinks it is because they stay calm and do not spend their energy moving around, so they can use all of their yolk sac just for putting on weight.

9. See Helmreich (2009) for a discussion of life-forms and forms of life.

10. This combination of mobility and durability was later named "immutable mobiles" (Latour 1987).

11. These tanks are about 150 centimeters in diameter and nearly a meter high. As the fry put on weight and become parr, they will be transferred to a building where the tanks are bigger.

12. The drain does not empty directly into the sewer but is connected to a filter, which is put in place as yet another physical boundary between the farmed and the wild, between inside and outside. The purpose of this filter is not so much to save each individual salmon (although that is occasionally done) as to protect the surrounding waterways from young farmed salmon finding their way out and eventually upriver, where they might interbreed with their distant cousins. (See also chapter 7.)

13. The light is turned on with a dimmer-switch because the sudden brightness when the light is turned on abruptly with a regular switch scares the fish and makes them crowd together at the bottom. (See also chapter 4, "Holding Tanks Together.")

14. Despret 2013; see also chapter 3.

15. This is called Specific Growth Rate (SGR) and is involved in calculating FCR.

16. When salmon smoltify, they lose their characteristic "parr marks" and become silvery; their cells undergo changes that enable them to maintain a mineral balance in water of higher salinity; and their buoyancy increases as they become more streamlined (Verspoor, Stradmeyer, and Nielsen 2007, 36).

17. According to Verspoor, Stradmeyer, and Nielsen (2007), "the 'priming' and 'releasing' factors linked to the initiation and maintenance of smolt migration are seasonal changes in photoperiod, water temperature, lunar phase, and water flow" (36).

18. While commercial salmon fisheries in the North Pacific rely on migrant workers to accommodate the seasonal nature of salmon migration, salmon farming in Norway has a much steadier labor requirement. In Norway, where labor unions are strong, permanent employment is generally expected, and most salmon producers rely primarily on permanent staff, usually from the local community.

19. According to the vet, those on a steered regime (the autumn smolts) appear to be doing very well—some would say better than the others—and she speculates that one less summer season may actually be a good thing, considering the stress associated with delousing, which is done less often in winter than in summer, when river salmon pass through the fjord.

20. Our findings are not unique to aquaculture. As Vicky Singleton shows in her study of cattle in the United Kingdom (2010, 250, 252), control of cattle movement on the farm is a precarious achievement. The aims of control inherent in the so-called Cattle Tracing System remain in tension and are crucially different from the farming practices that she encounters. While the former dream of control, the latter are crafted by care.

21. Pink and chum salmon are the two subspecies that are most successfully raised in Alaskan hatcheries, while sockeye, which is the prime species caught in Bristol Bay (cf. Hébert 2010), is less successful (Grant 2012).

22. In other words, their scalability is maintained; see Tsing 2012, 507.

6. BECOMING SENTIENT

1. In a passage that is essentially a pledge for greater legal protection of slaves, Bentham ([1789] 1907) wrote:

> The day *may* come, when the rest of the animal creation may acquire those rights which never could have been withholden from them but by the hand of tyranny. The French have already discovered that the blackness of the skin is no reason a human being should be abandoned without redress to the caprice of a tormentor. It may one day come to be recognised that the number of the legs, the villosity of the skin, or the termination of the os sacrum are reasons equally insufficient for abandoning a sensitive being to the same fate.... The question is not, Can they reason? nor, Can they talk? but, Can they suffer? (note 122)

2. That is, "a being with a complex mental life, including perception, desire, belief, memory and a sense of the future" (Regan 1983, cited in Lund et al. 2007, 3).

3. Anna Tsing proposes the term *worlding* for propositions about context, or what she defines as "the always experimental and partial, and often quite wrong, attribution of worldlike characteristics to scenes of social encounter" (2010, 47). In her analysis, worlding exercises serve as "figurings of relevant social worlds," and it is in this sense that I suggest *worlding* as a helpful term in capturing the ontological reclassification of fish as animals (Tsing 2010, 47–49).

4. Huntingford et al. (2006, 334) apply what they call a "feelings-based" approach to animal welfare, which focuses on the animals' subjective mental states. This approach is contrasted with "function-based" and "nature-based" approaches. The former centers on an animal's adaptive abilities and typically emphasizes good health, while the latter assumes that each animal has an inherent biological nature that it must express and sees good welfare as the fulfillment of such "natural" needs. In this brief review, I follow their cues and focus primarily on the first. For a critique, see Arlinghaus et al. (2007).

5. I draw on Singleton and Law (2012, 1–2), who define *devices* as "practices of purposive crafting" and note that although the word *devices* can sound "machine-like," devices may also be "purposeful plans or schemes" and they "can also take the form of words."

6. I am inspired by Charis Thompson's use of the term *choreographies* in the book *Making Parents* (2005), where she defines "ontological choreographies" as "the dynamic coordination of the technical, scientific, kinship, gender, emotional, legal, political and financial aspects of ART clinics. She describes these clinics as a "deftly balanced coming together of things that are generally considered part of different ontological orders" (9). In a similar move, John Law (2010) uses the term *choreography* as a way of understanding care in veterinary practices (67). See also Charis Cussins's (1996 [same author, different surname]) use of choreographies as an ontological and political metaphor "to invoke materiality, structural constraint, performativity, discipline, co-dependence of setting and performers, and movement" (cited in Law 2010, 67). When I speak of choreographies of caring and killing, I seek, in a similar way, to draw attention to caring and killing as multiple practices that involve people, machines, inscriptions, fish, and legal documents. Furthermore, I wish to emphasize that these practices enact specific orderings in and of time and space—or

spatiotemporal choreographies. Killing a salmon at a slaughterhouse, for instance, involves both knives and electrical voltage, but the temporal sequence in which they are used matters a great deal and enacts a spatial and social separation too. See also Vialles (1994), which makes a related point about cattle slaughter in France, as well as this chapter's section on slaughter, "Becoming Food: Notes from the Slaughterhouse."

7. That we searched for the undersized parr and finally found them is entirely due to the alertness of my fieldwork collaborator, John Law. Although we had spent an entire week together in the same vaccination room, I had not even noticed that some fish went missing. Clearly, the choreographies of killing succeeded in enacting a "new normal" and in making the substandard fish invisible to me. I thank John for being more perceptive, or perhaps more attuned to the presence of that which is suppressed in the everyday (see Law and Lien 2014 for John's elaboration of this point).

8. Practically all field visits were done in collaboration with John Law. Occasionally, Eira accompanied us as well, as a visiting student from Oslo among sixth and seventh graders at the local school.

9. We had vaccinated these fish three months earlier, when their movements had been quick and alert.

10. How many young salmon does it take to fill a bucket? The number had never been relevant before, but my coworker Kristin made an estimate and wrote the numbers on a sheet of paper: 220 to 180, depending on the size of the fish in each tank (during vaccination, the fish were sorted and redistributed according to size). Then all we had to do was count the buckets.

11. The vet explained that when it is cold, water crystallizes on the gills and may damage them.

12. As I emptied my buckets into the container, I thought about the earthquake in Haiti just a couple of weeks earlier. I had seen images of bulldozers clearing the soil to make mass graves for thousands of people. According to the news, 150,000 people had died. I could not imagine such a number of people until the day the manager said that we had now passed 150,000 dead fish. The container was two-thirds full. Each fish was about 20 centimeters long. Each bucket held approximately two hundred fish. Each tank produced several buckets of dead fish each day. Every week had many days, and the cold had lasted for many weeks already. I watched the massive pile of small, stiff, frozen fish and, for the first time, I was able to sense the scale of the Haitian catastrophe.

13. About a quarter of the production at this site is sent by air, mostly to East Asia. The rest is sent by trailers to continental Europe.

14. A similar situation is described by John Law (2010), who, in the context of veterinary practices, notes that choreographies of care depend on the organization and unfolding of separations. His example concerns the vet's care for the calf and his care for the self, which do not always go together and therefore involve "routines for separating moments and objects of care and the subjectivities that go with them" (68). Something similar may be going on with Arne. In both cases, separations unfold through the multiple practices of caring and killing.

15. These regulations state that: "Fish present substantial physiological differences from terrestrial animals and farmed fish are slaughtered and killed in a very different context, in particular as regards the inspection process. Furthermore, research on the stunning of fish

is far less developed than for other farmed species. Separate standards should be established on the protection of fish at killing. Therefore, provisions applicable to fish should, at present, be limited to the key principle" (303/2). In January 2015, no separate standards had yet been established within the EU. For the complete text of Council Regulation (EC) No. 1099/2009 of 24 September 2009 on the protection of animals at the time of killing, see http://ec.europa.eu/food/animal/welfare/slaughter/regulation_1099_2009_en.pdf.

16. See Ministry of Agriculture and Food, 2009, Animal Welfare Act *(Dyrevelferdsloven)*, https://www.regjeringen.no/en/dokumenter/animal-welfare-act/id571188/. The proposition to the Parliament (Ot. prp. nr. 15 [2008–2009] Om lov om dyrevelferd [About the Animal Welfare Act]) is available at https://www.regjeringen.no/nb/dokumenter/otprp-nr-15-2008-2009-/id537570/.

17. Forskrift om slakterier og tilvirkningsanlegg for akvakulturdyr [Regulations concerning slaughterhouses and processing plants for aquaculture], 2006, Nærings- og fiskeridepartementet [Ministry of Trade, Industry, and Fisheries], https://lovdata.no/dokument/SF/forskrift/2006-10-30-1250. The regulations were implemented 1 August 2008, but several dispensations regarding the use of CO_2 implied that they were not fully enforced until 1 July 2012.

Article 14 of Forskrift om slakterier og tilvirkningsanlegg for akvakulturdyr, about anesthetizing fish during slaughter, states that "Fish shall be anesthetized before or at the time of slaughter, and remain so until death. Anesthetization shall be done through a method that does not involve significant stress or pain. If necessary, the fish shall be sedated or immobilized before being anesthetized. It is illegal to anesthetize fish by using gas, including CO_2, or any other medium that blocks the oxygen uptake, including salt, ammonia or other chemicals with similar effects" (my translation). [Fisk skal bedøves før eller samtidig med avliving og være bedøvd til døden inntrer. Bedøving skal skje ved egnet metode som ikke påfører fisken vesentlig stress eller smerte. Om nødvendig skal fisken sederes eller immobiliseres på forsvarlig måte før bedøving. Det er forbudt å bedøve fisk ved hjelp av gass, herunder CO_2, eller annet som blokkerer oksygenopptaket, samt salt, salmiakk eller andre kjemikaler med lignende virkning.]

18. According to the act (§2132g Definitions), "[T]he term 'animal'" means any live or dead dog, cat, monkey (nonhuman primate mammal), guinea pig, hamster, rabbit, or such other warm-blooded animal, as the Secretary may determine is being used, or is intended for use, for research, testing, experimentation, or exhibition purposes, or as a pet; but such term excludes (1) birds, rats of the genus *Rattus,* and mice of the genus *Mus,* bred for use in research, (2) horses not used for research purposes, and (3) other farm animals, such as, but not limited to livestock or poultry, used or intended for use as food or fiber, or livestock or poultry used or intended for use for improving animal nutrition, breeding, management, or production efficiency, or for improving the quality of food or fiber." 7 U.S.C. 54, 2009 edition, Transportation, Sale, and Handling of Certain Animals, www.gpo.gov/fdsys/pkg/USCODE-2009-title7/html/USCODE-2009-title7-chap54.htm.

19. American Fisheries Society, *Guidelines for Use of Fish in Field Research,* Policy Statement no. 16, approved September 1987, www.nal.usda.gov/awic/pubs/Fishwelfare/AFS16.pdf.

20. In her analysis of the inclusion of mistreatment of animals in the Norwegian penal code from 1902, Asdal (2012, 395) shows that while the question of animal consciousness and animal pain remained unsettled, concern with mistreatment of animals was defined

within a moral framework and was justified in relation to a notion of decency. In other words, punishing mistreatment of animals was justified with reference to sensibilities of humans and society as a collective rather than to animal sentience.

21. Forskrift om slakterier og tilvirkningsanlegg for akvakulturdyr [Regulations concerning slaughterhouses and processing plants for aquaculture]. See note 17 for translation of Article 14.

22. I am grateful to Arturo Escobar for posing the question so succinctly, and to John Law, Marisol de la Cadena, and Mario Blaser for putting salmon on the agenda of the Sawyer seminar at University of California, Davis, in the fall of 2012.

7. BECOMING ALIEN

1. *Ufesk* (nonfish) is a common Norwegian term for fish that is undesirable or inedible.

2. The right to fish the river is traditionally distributed among local farmers who own the property adjacent to the river. This hereditary right of access can then be rented to others, such as tourists.

3. Prices vary and are settled at the fall auctions. For the 2014 season, an all-time record was set when two days of fishing in the river Alta was auctioned off to a British fisherman for 150,000 kroner, or US$25,000 (Lindi 2013).

4. This equals (calculating the numbers on the basis of the report from January 2012) a biomass of 672,183 metric tons (Directorate of Fisheries 2013a).

5. These estimates are based on the aquaculture industry's reports to authorities and other indications (e.g., local fishermen). Estimates are uncertain and some underreporting may be assumed.

6. Sunnset 2013.

7. When the "red list" of threatened species in Norway was published in 2010, salmon was not on it, but the report still recognized that the salmon stocks had diminished: "Norway has 452 large and small salmon rivers and watersheds located from Østfold in the South to Finnmark in the North. However, during the past decades there has been a population decline for Atlantic salmon. The species is now extinct in 45 rivers and is under a high risk of becoming extinct from another 32 rivers" [Norge har 452 store og små lakseførende vassdrag, som er lokalisert fra Østfold i sør til Finnmark i nord. Det har imidlertid gjennom flere tiår vært en negativ bestandsutvikling hos atlantisk laks. Arten er nå borte fra 45 elver og risikoen er vurdert til å være høy for at den også skal bli borte fra 32 andre elver.] (Kålås et al. 2010a).

8. Other significant threats include sea lice, acidic water, the parasite *Gyrodactylus salaris*, hybridization with sea trout *(sjøaure)*, various diseases, overfishing, and hydroelectric dams. Variation in ocean temperatures and the prevalence of prey are also likely to affect growth and survival during the sea phase (Kålås et al. 2010b, 407–8).

9. Around 1850, the regulation, cultivation, and control of salmon resources had become a significant national issue in Norway, and by the end of the nineteenth century, new hatcheries produced as many as a million fry per year (Treimo 2007).

10. The section "'Becoming Alien'; or, the Performance of Nature as a Place Where Humans Are Not" draws on Lien and Law 2011.

11. The rescue plan was based on a research program conducted from 2000 to 2007, and its first report, "Now or Never for the Vosso Salmon" [Nå eller aldri for Vossolaksen], was published by the Directorate of Nature Management in 2008 (Barlaup 2008).

12. The first of the rescue plan's two pillars involves intensive upscaling of the Vosso hatchery, in order to supply the river with genetically distinct Vosso salmon, something that has not been done on a similar scale in Norway before. The second involves a variety of measures, the most important of which aim to reduce smolt mortality due to sea salmon lice and to reduce the prevalence of escaped farmed salmon in the rivers. Additional measures were implemented to monitor water quality and mitigate the effects of the hydroelectric power infrastructure.

13. This understanding of wildness is expressed among fishermen and among some biologists too. I interviewed a biologist working with the project who put it this way:

> A wild fish is a fish that has grown in the river, on its own, hatched by wild fish, and then returned. A cultivated fish, however, is a fish that is the result either of our planting eggs in the river or producing fry and releasing them in the river. And then there are the fish we produce until they become smolts, which we release at different sites [along the watershed]. So in practice, there are three degrees of cultivated fish. When a cultivated fish returns as broodstock and hatches in the river, it becomes a philosophical question whether its fry [offspring] are cultivated or wild. (my translation) [En vill fisk er jo en fisk som er oppvokst på egenhånd i elven, gytt av villfisk og overlevd og kommet tilbake. Mens en kultivert fisk, det er fisk som er avkommet enten ved at vi har plantet eggene i elven, eller den kan ha vært produsert frem som yngel og sluppet i elven, også er det den vi produserer fram til smolt og slipper på forskjellige steder. Så i praksis er det tre grader av kultivert fisk. Så kan det jo bli et filosofisk spørsmål da, når den kultiverte fisken kommer tilbake som stamfisk og gyter i elven, om denne yngelen som kommer er kultivert eller vill.]

14. "The goal is that our increased cultivation efforts, in combination with measures to reduce the most likely threats, will provide the basis for a self-reproducing, viable, and sustainable [harvestable] stock" (my translation). [Målsettingen er at den økte kultiveringsinnsatsen, kombinert med tiltak for å redusere sannsynlige trusselfaktorer, skal gi grunnlaget for en selvreproduserende, livskraftig og høstbar bestand.] Barlaup 2008, 9.

15. Worn fins are often (but not always) seen in adult farmed salmon, the result of a life spent inside in a physical enclosure, where a salmon is likely to often touch the netting and other fish. Because worn fins are easily visible, they are commonly looked for to distinguish escaped farmed salmon from salmon that return from the ocean.

16. In addition, there are various subcategories of cultivated fish established through controlled experiments that test, for example, whether early treatment with a lice-medicine will make the salmon less susceptible to sea lice on their journey to sea. Salmon that are part of such experiments are tagged, so they also have to be cut open in order for the tag to be analyzed and properly categorized as data (Dalheim 2012, 91–92).

17. Comparing genetic profiles can reveal similarities and differences between current salmon caught in the river and those that should have been there. This comparison is about degrees of genetic overlap rather than absolute genetic distinctions. In the Vosso River, such

comparisons are possible because of the genetic stock that was removed from the river in the 1980s (Eidfjord), before aquaculture took off, and now serves as a kind of baseline for studying subsequent genetic alterations.

18. These annual meetings usually take place in December—I attended one in 2011, at Voss.

19. Even though classifications multiply and the various techniques often reveal a bewildering complexity—sometimes allowing what appears to me as some provisional noncoherence—the annual meeting (like other scientific practices, such as peer-reviewed journals) is structured around a common scientific aim. The aim is to produce not a multiple salmon but rather *a Vosso salmon* that is part of a specific, singular universe (rather than a pluriverse; see, for example, Escobar [2008]). It is a classic scientific strategy for establishing and maintaining what John Law (2014) calls a "one-world world," although it does not necessarily always succeed.

20. *Scaling devices* in both senses of the term: fish scales as well as measuring scales.

21. One plant located on the riverside is Toro, a major food manufacturer. Toro let the angler's association use a deserted factory building to cultivate fry for the river.

22. You have to have an Arna mailing address to be a member.

23. The website of the Arna anglers is at http://www.arnasportsfiskarlag.com/fangst statistikk/?__requestid=__requestid528a070b37239.

24. "We are aware that there is some doubt concerning the origin of the current stock in Storelva. One can nevertheless not exclude [the possibility] that part of the original stock has contributed to the creation of the genetic base [of the current stock]." [Vi er kjent med at det er tvil rundt opphavet til dagens stamme i Storelva. Ein kan i dag likevel ikkje uteluke at ein del av den opphavlege stamma har vore med og danna det genetiske grunnlaget.] Fylkesmannens brev (letter from the Hordaland County Governor) to the Arna Sportsfiskarlag (Arna Anglers Association), October 14, 2011, http://salmongroup.no/wp-content /uploads/2011/12/FeilvandringavVossolakstilStorelvaiArna-orientering.pdf.

25. "Meir føre var går det vel ikkje an å bli, enn å tømme ei elv for ein type villaks for å verne ein stamme som ingen har fastslått at eksisterar." Øyvind Kråkås, 2011, "Vil verne om stamme ein ikkje veit om eksisterar," Aktuelt, Salmon Group, http://salmongroup.no /aktuelt/2011/10/vil-verne-om-stamme-ein-ikkje-veit-om-eksisterar/. The Salmon Group is a network of locally owned salmon farms in Norway (see Salmon Group, Om Oss [About Us], http://salmongroup.no/om-oss/).

26. Website guidelines for 2013 instructed the Arna anglers to take fin-clipped salmon out of the river, take scale samples, and place the samples in the freezer of the Vosso hatchery, before they bring the salmon home: "Får du vossolaks skal den ikke registreres på dette skjemaet, men tas opp, ta skjellprøver, merk den vossolaks og legg den i fryseboksen på klekkeriet, (sammen med skjellprøven). Oppdrettslaksen kan du ta med hjem, etter at den er nøye registrert." Arna Sportsfiskarlag, accessed November 18, 2013, http://www .arnasportsfiskarlag.com/.

27. Another possibility is the failure of a pipeline during transport, which might accidentally send salmon into the sea rather than their destined tank or pen.

28. Do they search for food? Maybe. Do they have an inborn inclination to move upstream? Some probably do. Salmon farmers have noted that some even migrate towards the hatchery where they spent the first part of their life. Perhaps they are simply trying to get on with their life.

29. The Norse term *fredløs* is from the Viking era, when Norway was made Christian by the alignment of the Roman church with military power and when those who refused were chased from their farms and stripped of their positions and the security of their local community as well as their personal belongings. Many gathered in the mountains, where they led a precarious existence and could be killed by anyone without legal repercussions—hence the term *fredløs* (without peace). As *fredløs*, our salmon survivors may figure out a way to eat and perhaps even to reproduce, but their future is highly uncertain, and if the Vosso salmon rescue project gets its way, they will be a historical relic, like the medieval resistance fighters.

30. What is really the problem posed by the "escaped farmed salmon"? Is it their procreative potential as future ancestors of a hybridized stock? Or is it their affinity with the aquaculture industry, an industry that is associated (perhaps especially by anglers but also by many others) with "unnaturalness," industrial exploitation of nature, spoiled river habitats, and waste? The arguments tend to blur, and an attempt at keeping them separate seems unnecessary for the present analysis.

8. TAILS

1. Strathern 1991; see also chapter 1.
2. Barad 2003.
3. Comparison is, after all, based upon reification. Hence, as Tsing (2014) argues, drawing on Strathern: "reification to create comparison is useful if it serves critical reflections," provided that such reification is both serious and playful (223).
4. "Arts of Living on a Damaged Planet" was the subtitle for a conference on the Anthropocene at the University of California, Santa Cruz in May 2014. For details, see http://ihr.ucsc.edu/portfolio/anthropocene-arts-of-living-on-a-damaged-planet/.
5. Brit Ross Winthereik and Helen Verran (2012) describe this dual logic as a contrast between two kinds of ethnographic interventions. On the one hand, there are what they call "one-many generalizations," which are concerned with transforming space-time specificities into general claims, and on the other, "whole-parts generalizations" which recognize that the relations between the parts do not exist beforehand, but must be "crafted and bundled" together, hence the focus is rather on the generative potential of any representation, as well as its incompleteness (46, 47). Another, more conventional way of framing a related contrast is to say that it is a matter of choosing between a historical and a processual approach.
6. If a fisherman on an ocean trawler sees salmon differently, it is most likely not because he is less kind than Tone, or male, or because his cultural background resists the attribution of sentience to fish, but because the sociomaterial conditions and concerns of *his* workplace invite different sentiments. If you see thousands of fish crammed together in a purse seine, it is probably difficult to think of fish as sentient beings, just as it may be difficult for an aquaculture operation manager, hard-pressed to deliver enough salmon for slaughter, to spend a lot of time thinking about the implication of density on salmon welfare.
7. See Taranger et al. 2011. The risk assessment defined feed and feed resources as one of five relevant goals but did not include these considerations in the assessment report.

REFERENCES

Aarseth, Bernt, and Stig Erik Jakobsen. 2004. *On a Clear Day You Can See All the Way to Brussels: The Transformation of Aquaculture Regulation in Norway.* SNF Working Paper no. 63/04. Bergen: Institute for Research in Economics and Business Administration.

Abram, Simone, and Marianne E. Lien. 2011. "Performing Nature at the World's Ends." *Ethnos* 76 (1): 3–18.

Alvial, Adolfo, Frederick Kibenge, John Forster, José M. Burgos, Rolando Ibarra, and Sophie St.-Hilaire. 2012. *The Recovery of the Chilean Salmon Industry: The ISA Crisis and Its Consequences and Lessons.* Puerto Montt, Chile: Global Aquaculture Alliance. www.gaalliance.org/cmsAdmin/uploads/GAA_ISA-Report.pdf.

Anderson, David, Rob P. Wishart, and Virginie Vaté, eds. 2013. *About the Hearth: Perspectives on the Home, Hearth and Household in the Circumpolar North.* New York and Oxford: Berghahn.

Anderson, Virginia de John. 2006. *Creatures of Empire: How Domestic Animals Transformed Early America.* Oxford: Oxford University Press.

Anneberg, Inger. 2013. "Actions of and Interactions between Authorities and Livestock Farmers—in Relation to Animal Welfare." PhD diss., Science and Technology, Department of Animal Science, Aarhus University.

Aquagen. 2014. Products: Eyed Eggs of Atlantic Salmon *(Salmo salar L.).* http://aquagen.no/en/products/salmon-eggs/.

Ardener, Edwin. 1989. *The Voice of Prophecy and Other Essays.* Edited by Malcolm Chapman. London: Basil Blackwell.

Arlinghaus, Robert, Steven J. Cooke, Alexander Schwab, and Ian G. Cows. 2007. "Fish Welfare: A Challenge to the Feelings-Based Approach, with Implications for Recreational Fishing." *Fish and Fisheries* 8: 57–71.

Asche, Frank, Kristin H. Roll, and Sigbjørn Tveterås. 2007. "Markedsvekst som drivkraft for laksenæringen." In *Havbruk: Akvakultur på norsk,* edited by Bernt Aarseth and Grete Rusten, 51–69. Oslo: Fagbokforlaget.

Asdal, Kristin. 2012. "Contexts in Action—and the Future of the Past in STS." *Science, Technology & Human Values* 4: 379–403.

———. 2014. "Enacting Values from the Sea: On Innovation Devices, Value-Practices and the Co-Modifications of Markets and Bodies in Aquaculture." In *Value Practices in the Life Sciences and Medicine,* edited by Isabel Dussauge, Claes-Fredrik Helgesson, and Francis Lee. Oxford: Oxford University Press.

Barad, Karen. 2003. "Posthumanist Performativity: Towards an Understanding of How Matter Comes to Matter." *Signs: Journal of Women in Culture and Society* 28 (3): 801.

Barlaup, Bjørn T. 2008. "Nå eller aldri for Vossolaksen—anbefalte tiltak med bakgrunn i bestandsutvikling og trusselfaktorer." *DN utredning 2008-9.* Trondheim: Direktoratet for Naturforvaltning (Directorate of Nature Management).

———. 2013. "Redningsaksjonen for Vossolaksen." *DN utredning 2013.* Trondheim: Direktoratet for Naturforvaltning.

Barnes, John. 1954. "Class and Committees in a Norwegian Island Parish." *Human Relations* 7 (39).

Barrionuevo, Alexei. 2009. "Chile's Antibiotics Use on Salmon Farms Dwarfs That of a Top Rival's." *New York Times,* July 26. www.nytimes.com/2009/07/27/world/americas/27salmon.html?_r=0.

Bentham, Jeremy. (1789) 1907. "Of the Limits of the Penal Branch of Jurisprudence." Chap. 17 in *An Introduction to the Principles of Morals and Legislation.* Facsimile of the 1823 edition, Library of Economics and Liberty. www.econlib.org/library/Bentham/bnthPML18.html.

Berge, Aslak. 2005. *Salmon Fever: The History of Pan Fish.* Bergen, Norway: Octavian.

Berge, Dag Magne. 2002. "Dansen rundt gullfisken: Næringspolitikk og statlig regulering i norsk fiskeoppdrett 1970–1997." PhD diss., University of Bergen.

Boas, Franz. (1911) 1938. *The Mind of Primitive Man.* New York: Macmillan. Facsimile, Internet Archive. https://archive.org/details/mindofprimitivem031738mbp.

Bowker, C. Geoffrey, and Susan Leigh Star. 1999. *Sorting Things Out.* Cambridge, MA: MIT Press.

Brøgger, Benedicte. 2010. "An Innovative Approach to Employee Participation in a Norwegian Retail Chain." *Economic and Industrial Democracy* 31 (4): 477–95.

Bshary, Recouan, Wolfgang Wickler, and Hans Fricke. 2002. "Fish Cognition: A Primate's Eye View." *Animal Cognition* 5: 1–13.

Bubandt, Nils, and Ton Otto. 2010. "Anthropology and the Predicaments of Holism." In *Experiments in Holism,* edited by Ton Otto and Nils Bubandt, 1–17. Oxford: Blackwell.

Buller, Henry. 2013. "Individuation, the Mass, and Farm Animals." *Theory, Culture & Society* 30 (7/8): 154–75.

Buller, Henry, and Carol Morris. 2003. "Farm Animal Welfare: A New Repertoire of Nature-Society Relations or Modernism Re-embedded." *Sociologia Ruralis* 43 (3): 217–37.

Byrkjeflot, Haldor. 2001. "The Nordic Model of Democracy and Management." In *The Democratic Challenge to Capitalism,* edited by Haldor Byrkjeflot, Sissel Myklebust, Christine Myrvang, and Francis Sejersted, 19–45. Bergen: Fagbokforlaget.

Cabello, Felipe C., Henry P. Godfrey, Alexandra Tomova, Larissa Ivanova, Humberto Döltz, Ana Millanao, and Alejandro Buschman, H. 2013. "Antimicrobial Use in Aquaculture Re-examined: Its Relevance to Antimicrobial Resistance and to Animal and Human Health." *Environmental Microbiology*: 1–26. doi:10.1111/1462-2920.12134.

Cadena, Marisol de la. 2014. "The Politics of Modern Politics Meets Ethnographies of Excess through Ontological Openings." *Cultural Anthropology Online*. http://culanth.org/fieldsights/471-the-politics-of-modern-politics-meets-ethnographies-of-excess-through-ontological-openings.

Candea, Matei. 2010. "'I Fell in Love with Carlos the Meerkat': Engagement and Detachment in Human-Animal Relations." *American Ethnologist* 37 (2): 241–58.

Cassidy, Rebecca. 2007. "Introduction: Domestication Reconsidered." In *Where the Wild Things Are Now: Domestication Reconsidered*, edited by Rebecca Cassidy and Molly Mullin, 1–27. Oxford: Berg.

Cassidy, Rebecca, and Molly Mullin. 2007. *Where the Wild Things Are Now: Domestication Reconsidered*. Oxford: Berg.

Chandroo, K. P., I. J. H. Duncan, and R. D. Moccia. 2004. "Can Fish Suffer? Perspectives on Sentience, Pain, Fear and Stress." *Applied Animal Behaviour Science* 86: 225–50.

Childe, V. G. 1958. "Retrospect." *Antiquity* 32: 69–74.

Childe, V. G., and G. Clark. 1946. *What Happened in History?* New York: Penguin.

Chutko, Per Ivar. 2011. "En temmelig vill en: Kontroverser om laks, ca. 1880–2009." Master's thesis, Department of Interdisciplinary Studies of Culture, Norwegian University of Science and Technology (Norges teknisk-naturvitenskapelige universitet [NTNU]).

Clastres, Pierre. 1977 (1974). *Society against the State*. Translated by Robert Hurley. New York: Urizen Books.

Clutton-Brock, Juliet. 1994. "The Unnatural World: Behavioral Aspects of Humans and Animals in the Process of Domestication." In *Animals and Human Society*, edited by A. Manning and J. A. Serpell, 23–36. London: Routledge.

Council of Europe, Standing Committee of the European Convention for the Protection of Animals Kept for Farming Purposes (T-AP). 2005. "Recommendations concerning Farmed Fish." Adopted on December 5, 2005. www.coe.int/t/e/legal_affairs/legal_co-operation/biological_safety_and_use_of_animals/farming/Rec%20fish%20E.asp#TopOfPage.

Council of the European Union. 2009. "Council Regulation (EC) No. 1099/2009 of 24 September 2009 on the Protection of Animals at the Time of Killing." http://eur-lex.europa.eu/legal-content/EN/TXT/?uri=celex:32009R1099.

Cussins, Charis. 1996. "Ontological Choreography: Agency through Objectification in Infertility Clinics." *Social Studies of Science* 26 (3): 575–610.

Dalheim, Line. 2012. "Into the Wild and Back Again: Hatching 'Wild Salmon' in Western Norway." Master's thesis, University of Oslo.

Damsgaard, Børge. 2005. *Ethical Quality and Welfare in Farmed Fish*. European Aquaculture Society Special Publication, 35: 28–32.

Descola, Philippe. 2012. "Beyond Nature and Culture: Forms of Attachment." *HAU Journal of Ethnographic Theory* 2 (1): 447–78.

Despret, Vinciane. 2013. "Responding Bodies and Partial Affinities in Human-Animal Worlds." *Theory, Culture & Society* 30 (7/8): 51–76.

Directorate of Fisheries *(Fiskeridirecktoratet)*. 2009. "Om statistikken—Biomassestatistikk 4.1." Statistikk: Aqvacultur. Published November 26, 2009. www.fiskeridir.no/statistikk /akvakultur/om-statistikken/om-statistikken-biomassestatistikk.

———. 2013a. "Biomassestatistikk." Statistikk: Aqvacultur. Published November 26, 2009, accessed November 12, 2013. www.fiskeridir.no/statistikk/akvakultur/biomassestatistikk /biomassestatistikk.

———. 2013b. "Oppdaterte rømmingstall" [Updated escape figures]. Statistikk: Aqvacultur. Published March 18, 2005, accessed November 12, 2013. www.fiskeridir.no/statistikk /akvakultur/oppdaterte-roemmingstall.

Douglas, Mary. 1966. *Purity and Danger*. London: Routledge.

EFSA (European Food Safety Authority). 2004. "Opinion of the Scientific Panel on Animal Health and Welfare (AHAW) on a Request from the Commission related to Welfare Aspects of the Main Systems of Stunning and Killing the Main Commercial Species of Animals," EFSA-Q-2003–093. *EFSA Journal*. Adopted by the AHAW Panel June 15, 2004, published July 6, 2004, last updated October 11, 2004. doi:10.2903/j.efsa.2004.45. www.efsa.europa.eu/en/efsajournal/pub/45.htm.

———. 2009. "Species-Specific Welfare Aspects of the Main Systems of Stunning and Killing of Farmed Atlantic Salmon," EFSA-Q-2006–437. *EFSA Journal*. Adopted March 20, 2009, published April 14, 2009, last updated November 26, 2009. doi:10.2903 /j.efsa.2009.1011. www.efsa.europa.eu/en/scdocs/doc/1011.pdf.

Erbs, Stefan. 2011. "Writing the Blue Revolution: Theoretical Contributions toward a Contemporary History of Aquaculture." Master's thesis, Centre for Development and Environment, University of Oslo.

Escobar, Arturo. 2008. *Territories of Difference*. Durham, NC: Duke University Press.

Espeland, S. H., et al. 2010. *Kunnskapstatus leppefisk—Utfordringer i et økende fiskeri* [Current knowledge on wrasse: Challenge in an increasing fishery]. Bergen: Institute of Marine Research.

Evans-Pritchard, Edward E. 1964. *The Nuer: A Description of the Modes of Livelihood and Political Institutions of a Nilotic People*. New York and Oxford: Oxford University Press. Original edition, 1940.

FAO. 2009. "Part 1: World Review of Fisheries and Aquaculture." In *The State of the World Fisheries and Aquaculture 2008*. Rome: Food and Agriculture Organization of the United Nations, Fisheries and Aquaculture Department. www.fao.org/3/a-i0250e.pdf.

———. 2012. *The State of World Fisheries and Aquaculture 2012*. Rome: Food and Agriculture Organization of the United Nations. www.fao.org/docrep/016/i2727e/i2727e00.htm.

FHL (Fiskeri- og havbruksnæringens landsforening [Norwegian Seafood Federation]). 2011. *Norsk havbruk* [Aquaculture in Norway], August. www.fhl.no/getfile.php/ DOKUMENTER/eff_fhl_komplett_lowres.pdf (in English: http://fhl.no/wp-content/ uploads/importedfiles/Aquaculture%2520in%2520Norway%25202011.pdf).

Fish Pool ASA. 2014. "Price Information: Spot Prices." Accessed October 6, 2014. http:// fishpool.eu/spot.aspx?pageId=55.

Gad, Christian, Caspar Bruun Jensen, and Brit Ross Winthereik. 2015 (forthcoming). "Practical Ontology: Worlds in STS and Anthropology." *NatureCulture* 3.

Gederaas, Lisbeth, Ingrid Salvesen, and Åslaug Viken. 2007. *Norwegian Black List: Ecological Risk Analysis of Alien Species*. Trondheim: Norwegian Biodiversity Information Centre (Artsdatabanken). www.artsdatabanken.no/File/681/Norsk%20svarteliste%202007.

Gifford-Gonzalez, Diane, and Olivier Hanotte. 2011. "Domesticating Animals in Africa: Implications of Genetic and Archaeological Findings." *Journal of World Prehistory* 24 (1–23).

Godelier, Maurice. 1986. *The Mental and the Material*. London: Verso.

Grant, W. Stewart. 2012. "Understanding the Adaptive Consequences of Hatchery-Wild Interactions in Alaska Salmon." *Environmental Biology of Fishes* 94: 325–42.

Grasseni, Christina. 2004. "Skilled Visions: An Apprenticeship in Breeding Aesthetics." *Social Anthropology* 12: 41–57.

Gross, M. R. 1998. "One Species with Two Biologies: Atlantic Salmon *(Salmo salar)* in the Wild and in Aquaculture." *Canadian Journal of Fisheries and Aquatic Sciences* 55 (1): 131–44.

Gullestad, Marianne. 1992. *The Art of Social Relations: Essays on Culture, Social Action, and Everyday Life in Modern Norway*. Oslo: Norwegian University Press.

Gupta, Akhil, and James Ferguson. 1997. *Anthropological Locations*. Berkeley: University of California Press.

Haraway, Donna J. 1988. "'Situated Knowledges': The Science Question in Feminism and the Privilege of Partial Perspective." *Feminist Studies* 13 (3): 579–99.

———. 2008. *When Species Meet*. Minneapolis: University of Minnesota Press.

Hard, Jeffrey J., Mart R. Gross, Mikko Heino, Ray Hilborn, Robert G. Kope, RIchard Law, and John D. Reynolds. 2008. "Evolutionary Consequences of Fishing and Their Implications for Salmon." *Evolutionary Applications* 1 (2): 388–408. doi: 10.1111/j.1752-4571.2008.00020.x. http://onlinelibrary.wiley.com/doi/10.1111/j.1752-4571.2008.00020.x/full.

Harvey, Penny, and Hannah Knox. 2014. "Objects and Materials: An introduction." In *Objects and Materials,* edited by Penny Harvey, Eleanor Conlin Casella, Gillian Evans, Hannah Knox, Christine McLean, Elizabeth B. Silva, Nicholas Thoburn, and Kath Woodward, 1–19. London: Routledge.

Hastrup, Frida. 2011. "Shady Plantations: Theorizing Coastal Shelter in Tamil Nadu." *Anthropological Theory* 11 (5): 425–39.

Hastrup, Kirsten. 1992. *Det antropologiske prosjekt. Om forbløffelse*. København: Gyldendal.

Hébert, Karen. 2010. "In Pursuit of Singular Salmon: Paradoxes of Sustainability and the Quality Commodity." *Science as Culture* 19 (4): 553–81.

Helmreich, Stefan. 2009. *Alien Ocean: Anthropological Voyages in Microbial Seas*. Berkeley: University of California Press.

History Channel. 2012. *Mankind: The Story of Us All,* Series 1. Website. www.history.co.uk/shows/mankind-the-story-of-all-of-us/episode-guide/mankind-the-story-of-all-of-us-series-1.

Hodder, Ian. 1990. *The Domestication of Europe: Structure and Contingency in Neolithic Societies*. Oxford: Blackwell.

Howell, Signe. 2011. "Whatever Happened to the Spirit of Adventure?" In *The End of Anthropology?* edited by Holger Jebens and Karl-Heinz Kohl, 139–55. Wantage, Herefordshire, UK: Sean Kingston Publishing.

Huntingford, F., C. Adams, V. A. Braithwaite, S. Kadri, T. G. Pottinger, and P. Sandøe. 2006. "Current Issues in Fish Welfare." *Journal of Fish Biology* 68: 332–72.

Huntingford, Felicity. 2004. "Implications of Domestication and Rearing Conditions for the Behaviour of Cultivated Fishes." *Journal of Fish Biology* 65: 122–42.

Industry Standards for Fish. 1999. "Norwegian Industry Standard for Fish: Quality Grading of Farmed Salmon," NBS 10-01, version 2. http://fhl.nspo1cp.nhosp.no/files/Quality_grading_of_farmed_salmon.pdf.

Ingold, Tim. 1984. "Time, Social Relationships and the Exploitation of Animals: Anthropological Reflections on Prehistory." In *Animals and Archaeology*. Vol. 3, *Early Herders and Their Flocks*, edited by Juliet Clutton-Brock and C. Grigson, 3–12. Oxford: British Archaeological Reports.

———. 2000. *The Perception of the Environment*. London: Routledge.

———. 2011. *Being Alive: Essays on Knowledge and Description*. Oxford: Routledge.

———. 2013. "Prospect." In *Biosocial Becomings: Integrating Social and Biological Anthropology*, edited by Tim Ingold and Gísli Pálsson, 1–22. Cambridge: Cambridge University Press.

Ingold, Tim, and Gísli Pálsson. 2013. *Biosocial Becomings: Integrating Social and Biological Anthropology*. Cambridge: Cambridge University Press.

Institute of Marine Research (IMR; Havforskningsinstituttet). 2007. "Escaped farmed salmon is not an alien species" *[Rømt oppdrettslaks er ikke en fremmed art]*. Published May 31, 2007. www.imr.no/nyhetsarkiv/2007/mai/romt_oppdrettslaks_ikke_fremmedart/nb-no.

Jonsson, Bror, and Nina Jonsson. 2002. "Feilvandring hos laks." *Naturen* 6: 275–80.

Kålås, Jon Atle, Aslaug Viken, Snorre Henriksen, and Sigrun Skjelseth. 2010a. "LC: *Salmo salar*." *Norsk rødliste for arter 2010*. Trondheim: Artsdatabanken. www.artsportalen.artsdatabanken.no/#/Rodliste2010/Vurdering/Salmo+salar/25171.

———. 2010b. *Norsk rødliste for arter 2010* [2010 Norwegian red list for species]. Trondheim: Artsdatabanken (Norwegian Biodiversity Information Center). www.artsdatabanken.no/File/685/Norsk%20r%C3%B8dliste%20for%20arter%202010.

Kirksey, Eben, and Stefan Helmreich. 2010. "The Emergence of Multispecies Ethnography." *Cultural Anthropology* 25 (3): 545–76.

Kjærnes, Unni, Mark Harvey, and Alan Warde, eds. 2007. *Trust in Food: A Comparative and Institutional Analysis*. London: Palgrave Macmillan.

Knorr-Cetina, Karen. 1999. *Epistemic Cultures: How the Sciences Make Knowledge*. Cambridge, MA: Harvard University Press.

Kohn, Eduardo. 2013. *How Forests Think: Towards an Anthropology Beyond the Human*. Berkeley: University of California Press.

Kontali Analyse A/S. 2007. *The Salmon Farming Industry in Norway, 2007: Analysis of Annual Reports for 2006*. www.kontali.no/%5Cpublic_files%5Cdocs%5CThe_salmon_farming_industry_in_Norway_2007.pdf.

Larson, Greger, and Dorian Q. Fuller. 2014. The Evolution of Animal Domestication. *Annual Review of Ecology, Evolution and Systematics*. 45: 115–36.

Latimer, Joanna, and Mara Miele. 2013. "Naturecultures? Science, Affect and the Nonhuman." *Theory, Culture & Society* 30 (7/8): 5–31.

Latour, Bruno. 1987. *Science in Action*. Cambridge, MA: Harvard University Press.

———. 2005. *Reassembling the Social: An Introduction to Actor-Network Theory*. Oxford: Oxford University Press.

———. 2010. *On the Modern Cult of Factish Gods*. Durham, NC: Duke University Press.

Law, John. 1986. "On the Methods of Long Distance Control: Vessels, Navigation, and the Portuguese Route to India." In *Power, Action and Belief: A New Sociology of Knowledge? Sociological Review Monograph* 32, edited by John Law, 234–63. Henley, Oxfordshire, UK: Routledge.

———. 2002. *Aircraft Stories: Decentering the Object in Technoscience*. Durham, NC: Duke University Press.

———. 2004. "And If the Global Were Small and Noncoherent? Method, Complexity, and the Baroque." *Environment and Planning (D): Society and Space* 22: 13–26.

———. 2010. "Care and Killing: Tensions in Veterinary Practice." In *Care in Practice: On Tinkering in Clinics, Homes and Farms*, edited by Annemarie Mol, Ingunn Moser, and Jeannette Pols, 57–73. Bielefeld, Germany: Transcript Verlag.

———. 2015 (forthcoming). "What's Wrong with a One-World World?" *Distinktion: Scandinavian Journal of Social Theory*. (An earlier version, a paper presented to the Center for the Humanities, Wesleyan University, Middletown, CT, 19 September 2011, was published online by heterogeneities.net, September 25, 2011, www.heterogeneities.net/publications/Law2011WhatsWrongWithAOneWorldWorld.pdf.)

Law, John, Geir Afdal, Kristin Asdal, Wen-yuan Lin, Ingunn Moser, and Vicky Singleton. 2014. "Modes of Syncretism: Notes on Non-coherence." *Common Knowledge* 20 (1): 172–92.

Law, John, and Marianne E. Lien. 2013. "Slippery: Field Notes on Empirical Ontology." *Social Studies of Science* 43 (3): 363–78.

———. 2014. "Animal Architextures." In *Objects and Materials*, edited by Penny Harvey, Eleanor C. Casella, Gillian Evans, Hanna Knox, Christine McLean, Elizabeth B. Silva, Nicholas Thoburn, and Kath Woodward. New York: Routledge.

Law, John, and Vicky Singleton. 2012. "ANT and politics: Working in and on the world." www.sv.uio.no/sai/english/research/projects/newcomers/publications/working-papers-web/ant-and-politics.pdf.

Leach, Edmund. 1964. "Anthropological Aspects of Language: Animal Categories and Verbal Abuse." In *New Directions in the Study of Language*, edited by E. H. Lenneberg, 23–63. Cambridge, MA: MIT Press.

Leach, Helen N. 2003. "Human Domestication Reconsidered." *Current Anthropology* 44 (3): 349–68.

———. 2007. "Selection and the Unforeseeen Consequences of Domestication." In *Where the Wild Things Are Now: Domestication Reconsidered*, edited by Rebecca Cassidy and Molly Mullin, 71–101. Oxford: Berg.

Lévi-Strauss, Claude. 1966. *The Savage Mind*. Chicago: Chicago University Press.

Lien, Marianne E. 1997. *Marketing and Modernity*. Oxford: Berg.

———. 2005. "'King of Fish' or Feral Peril: Tasmanian Atlantic Salmon and the Politics of Belonging." *Society and Space* 23 (5): 659–73.

———. 2007a. "Domestication 'Downunder': Atlantic Salmon Farming in Tasmania." In *Where the Wild Things Are Now; Domestication Reconsidered*, edited by Rebecca Cassidy and Molly Mullin, 205–229. Oxford: Berg.

———. 2007b. "Feeding Fish Efficiently: Mobilising Knowledge in Tasmanian Salmon Farming." *Social Anthropology* 15 (2): 169–85.

———. 2009. "Standards, Science and Scale: The Case of Tasmanian Atlantic Salmon." In *The Globalization of Food*, edited by David Inglis and Debra Grimlin, 65–81. Oxford: Berg.

———. 2012. "Conclusion: Salmon Trajectories along the North Pacific Rim: Diversity, Exchange, and Human-Animal Relations." In *Keystone Nations. Indigenous Peoples and Salmon across the North Pacific*, edited by Benedict J. Colombi and James F. Brooks, 237–54. Santa Fe, NM: SAR Press.

Lien, Marianne E., and Aidan Davison. 2010. "Roots, Rupture and Remembrance; The Tasmanian Lives of Monterey Pine." *Journal of Material Culture* 15: 233–53.

Lien, Marianne E., and John Law. 2011. "'Emergent Aliens': On Salmon, Nature and Their Enactment." *Ethnos* 76 (1): 65–87.

Lien, Marianne E., Hilde Lidén, and Halvard Vike. 2001. *Likhetens Paradokser*. Oslo: Norwegian University Press.

Lindi, Marti. 2013. "150.000 kroner for to fiskedøgn i Altaelva," *Nordnytt*, NRK.no. October 25. www.nrk.no/nordnytt/150.000-kr-for-to-fiskedogn-i-alta-1.11318567

Lund, V., C. Mejdell, H. Röcklinsberg, R. Anthony, and T. Håstein. 2007. "Expanding the Moral Circle: Farmed Fish as Objects of Moral Concern." *Diseases of Aquatic Organisms* 75 (2): 109–18.

Magnussøn, Anne. 2010. "Making Food: Enactment and Communication of Knowledge in Salmon Aquaculture." PhD diss., University of Oslo.

Marx, Karl. (1844) 1961. *Economic and Philosophic Manuscripts of 1844*. Moscow: Foreign Languages Publishing House.

Mejdell, C, U. Erikson, E. Slinde, and K. Ø. Midling. 2010. "Bedøvingsmetoder ved slakting av laksefisk." *Norsk Veterinærtidsskrift* 122 (1): 83–90.

Menzies, Charles R. 2012. "The Disturbed Environment: The Indigenous Cultivation of Salmon." In *Keystone Nations: Indigenous Peoples and Salmon across the North Pacific*, edited by Benedict J. Colombi and James F. Brooks, 161–83. Santa Fe: SAR Press.

Midgley, Mary. 1983. *Animals and Why They Matter*. Athens: University of Georgia Press.

Mol, Annemarie. 1999. "Ontological Politics: A Word and Some Questions." In *Actor Network Theory and After*, edited by John Law and John Hassard, 74–89. Oxford and Keele: Blackwell and *Sociological Review*.

———. 2002. *The Body Multiple: Ontology in Medical Practice*. Durham, NC: Duke University Press.

———. 2008. *The Logic of Care: Care and the Problem of Patient Choice*. London: Routledge.

Mol, Annemarie, Ingunn Moser, and Jeannette Pols. 2010. *Care in Practice: On Tinkering in Clinics, Homes and Farms*, edited by Annemarie Mol, Ingunn Moser, and Jeannette Pols. Bielefeld, Germany: Transcript Verlag.

———. 2010. "Care: Putting Practice into Theory." In *Care in Practice: On Tinkering in Clinics, Homes and Farms*, edited by Annemarie Mol, Ingunn Moser, and Jeannette Pols, 7–27. Bielefeld, Germany: Transcript Verlag. www.transcript-verlag.de/ts1447/ts1447_1.pdf.

Nærings- og fiskeridepartementet (Ministry of Trade, Industries, and Fisheries). 2013. Verdens fremste sjømatnasjon [World's leading seafood nation]. Stortingsmelding [White paper]. https://www.regjeringen.no/nb/dokumenter/meld-st-22–20122013/id718631/.

Nash, Colin E. 2011. *The History of Aquaculture*. United States Aquaculture series. Ames, IA: Wiley-Blackwell.

Naylor, Rosamund, Ronald W Hardy, Dominique P. Bureaus, Alice Chiu, Matthew Elliott, Anthony P. Farrell, Ian Forster, Delbert M. Gatlin, Rebecca J. Goldburg, Katheline Hua, Peter D. Nichols, and Thomas F. Malone. 2009. "Feeding Aquaculture in an Era of Finite Resources." *Proceedings of the National Academy of Sciences in the United States of America* 106 (36): 15103–10.

NENT (Nasjonale Forskningsetiske komitee for naturvitenskap og teknologi [National Committee for Research Ethics in Science and Technology]). 1993. *Oppdrettslaks—en studie i norsk teknologiutvikling* (Farmed salmon—a study of Norwegian technology development). Oslo: NENT.

NOAA (National Oceanic and Atmospheric Administration), Northeast Fisheries Science Center. 2011. "Fish FAQ." June 6. www.nefsc.noaa.gov/faq/fishfaq1a.html.

Nordeide, Anita. 2012. "Møte mellom menneske og laks. Om laksepraksiser ved Namsenvassdraget." Master's thesis, University of Oslo. https://www.duo.uio.no/handle/10852/16273.

Norwegian Environment Agency (Miljøstatus). 2012. "Laks" [Salmon]. Ferskvann [Freshwater]. Published November 28, 2012. www.miljostatus.no/Tema/Ferskvann/Laks/#A.

Norwegian Food Safety Authority (Statens tilsyn for planter, fisk, dyr og næringsmidler). 2014. "Lakselusmiddelforbruket økte også i 2013." www.mattilsynet.no/fisk_og_akvakultur/fiskehelse/lakselusmiddelforbruket_okte_ogsaa_i_2013.12980.

Nustad, K., R. Flikke, and C. Berg. 2010. "Imagining Fish and Rivers in Aurland Norway." Paper presented to 11th Biennal European Association of Social Anthropologists (EASA) Conference, Maynooth, Ireland.

Nyquist, Jon Rasmus. 2013. "Making and Breaking the Invasive Cane Toad: Community Engagement and Interspecies Entanglements in the Kimberley, Australia." Master's thesis, University of Oslo.

Organization for Economic Co-operation and Development (OECD). 2012. "Closing the Gender Gap: Norway." Country Notes. *Closing the Gender Gap: Act Now.* www.oecd.org/norway/Closing%20the%20Gender%20Gap%20-%20Norway%20EN.pdf.

Osland, Erna. 1990. *Bruke havet . . . Pioner i norsk fiskeoppdrett.* Oslo: Det Norske Samlaget.

Otterå, Håkon, and Ove Skilbrei 2012. "Akustisk overvaking av seiens vandring I Ryfylkebassenget." *Rapport frå Havforskningen,* no. 14. April. Bergen: Institute of Marine Research.

Pálsson, Gísli. 1991. *Coastal Economies, Cultural Accounts: Human Ecology and Icelandic Discourse.* Manchester, UK: Manchester University Press.

Paxson, Heather. 2013. *The Life of Cheese: Crafting Food and Value in America.* Berkeley: University of California Press.

Pottage, Alain. 2004. "Introduction: The Fabrication of Persons and Things." In *Law, Anthropology and the Constitution of the Social: Making Persons and Things,* edited by A. Pottage and M. Mundy, 1–39. Cambridge: Cambridge University Press.

Reedy-Maschner, Katherine L. 2011. *Aleut Identities: Tradition and Modernity in an Indigenous Fishery.* Montreal, Quebec, and Kingston, Ontario: McGill–Queen's University Press.

Regan, Tom. 1983. *The Case for Animal Rights.* Berkeley: University of California Press.

Remme, Jon Henrik, Z. 2014. *Pigs and Persons in the Philippines: Human-Animal Entanglements in Ifugao Rituals.* Plymouth: Lexington Books.

Research Council of Norway. 2011. *Work Programme for the HAVBRUK Programme.* www.forskningsradet.no/servlet/Satellite?blobcol=urldata&blobheader=application%2Fpdf&blobheadername1=Content-Disposition%3A&blobheadervalue1=+attachment%3B+

filename%3D%22HAVBRUKProgramplaneng2011.pdf%22&blobkey=id&blobtable= MungoBlobs&blobwhere=1274505282241&ssbinary=true.

Rollin, Bernard E. 1995. *Farm Animal Welfare.* Ames: Iowa State University.

Rose, James D. 2002. "The Neurobehavioral Nature of Fishes, and the Question of Awareness and Pain." *Reviews in Fisheries Science* 10 (1): 1–38.

Russel, Nerina. 2002. "The Wild Side of Animal Domestication." *Society and Animals* 10 (3): 285–302.

———. 2007. "The Domestication of Anthropology." In *Where the Wild Things Are Now: Domestication Reconsidered*, edited by Rebecca Cassidy and Molly Mullin, 27–49. Oxford: Berg.

Scott, James. 2011. "Four Domestications: Fire, Plants, Animals, and . . . Us." Paper presented at the Tanner Lectures on Human Values, Harvard University, Cambridge, MA, May 4–6, 2011. http://tannerlectures.utah.edu/_documents/a-to-z/s/Scott_11.pdf.

Scott, Michael W. 2007. *The Severed Snake.* Durham, NC: Carolina Academic Press.

Singer, Peter. 1981. *The Expanding Circle: Ethics and Sociobiology.* New York: Farrar, Straus & Giroux.

Singleton, Vicky. 2010. "Good Farming: Control or Care?" In *Care in Practice: On Tinkering in Clinics, Homes and Farms*, edited by Annemarie Mol, Ingunn Moser and Jeannette Pols. Bielefeld: Transcript Verlag.

Singleton, Vicky, and John Law. 2012. "Devices as Rituals." *Journal of Cultural Economy* 6 (3): 259–77. www.tandfonline.com/doi/abs/10.1080/17530350.2012.754365#.VNnqK3bKxaQ.

Smith, Bruce. 2001. "Low-Level Food Production." *Journal of Archeological Research* (9): 1–43.

Solhaug, Trygve. 1983. *De norske fiskeriers historie 1815–1880.* 2nd ed. Bergen: Universitetsforlaget.

Star, Susan Leigh, and James Griesemer. 1989. "Institutional Ecology, 'Translations' and Boundary Objects: Amateurs and Professionals in Berkeley's Museum of Vertebrate Zoology, 1907–39." *Social Studies of Science* 19 (3): 387–420.

Statistics Norway. 2012. Historisk Statistikk: Fiskeoppdrett [Aquaculture]. Tabell 15.11: Anlegg med slakt av matfisk og slaktet mengde. www.ssb.no/a/histstat/tabeller/15-11.html#.

———. 2014. Fiskeoppdrett, 2013. Endelige tall [Final figures]. October 30. https://www.ssb.no/jord-skog-jakt-og-fiskeri/statistikker/fiskeoppdrett.

Stead, Selina M., and Lindsay Laird. 2002. *Handbook in Salmon Farming.* Chichester, West Sussex, UK: Praxis.

Stengers, Isabelle. 2011. "Another Science Is Possible! A Plea for Slow Science." Lecture at the Faculté de Philosophie et Lettres, Free University of Brussels, December 13, 2011. http://threerottenpotatoes.files.wordpress.com/2011/06/stengers2011_pleaslowscience.pdf.

Strathern, Marilyn. 1991. *Partial Connections.* Walnut Creek, CA: AltaMira Press.

———. 1992. *Reproducing the Future: Essays on Anthropology, Kinship and the New Reproductive Technologies.* Manchester, UK: Manchester University Press.

———. 2006. *Kinship, Law, and the Unexpected: Relatives Are Always a Surprise.* Cambridge: Cambridge University Press.

Sunnset, Beate Hoddevik. 2013. "Escapees Change the Wild Salmon." Institute of Marine Research (Havforskningsinstituttet). September 3. www.imr.no/nyhetsarkiv/2013/september/romt_oppdrettslaks_forandrer_villaksen/en.

Swanson, Heather Anne. 2013. "Caught in Comparison: Japanese Salmon in an Uneven World." PhD diss., Department of Anthropology, University of California, Santa Cruz.

———. 2015. "Shadow ecologies of conservation: Co-production of salmon landscapes in Hokkaido, Japan, and southern Chile." *Geoforum* 61 (2015): 101-110.

Taranger, Geir Lasse, Terje Svåsand, Abdullah S. Madhun, and Karin H. Boxaspen. 2011. "Risikoverdering—miljvirkninger av norsk fiskeoppdrett" [Risk assessment—environmental impacts of Norwegian aquaculture]. In *Fisken og Havet*. Bergen: Institute of Marine Research. (English translation at www.imr.no/filarkiv/2011/08/risk_assessment_engelsk_versjon.pdf/en.)

Thompson, Charis. 2005. *Making Parents: The Ontological Choreography of Reproductive Technology*. Cambridge, MA: MIT University Press.

Treimo, Henrik. 2007. "Laks, kart og mening. Det store laksegenomsprosjektet." PhD diss., University of Oslo.

Trigger, Bruce G. 1980. *Gordon Childe; Revolutions in Archaeology*. London: Thames and Hudson.

———. 1996. *A History of Archeological Thought*. Oxford: Oxford University Press.

Tsing, Anna. 2010. "Worlding the Matsutake Diaspora. Or, Can Actor-Network Theory Experiment with Holism?" In *Experiments in Holism*, edited by Ton Otto and Nils Bubandt, 47–66. London: Blackwell.

———. 2012a. "On Nonscalability: The Living World Is Not Amenable to Precision-Nested Scales." *Common Knowledge* 18 (3): 505–24.

———. 2012b. "Unruly Edges: Mushrooms as Companion Species." *Environmental Humanities* 1: 141, 154.

———. 2013. "More-Than-Human Sociality. A Call for Critical Description." In *Anthropology and Nature*, edited by Kristen Hastrup, 27–42. London and New York: Routledge.

———. 2014. "Strathern beyond the Human: Testimony of a Spore." *Theory, Culture and Society* 31 (2/3): 221–41.

———. 2015. *The Mushroom at the End of the World: On the Possibility of Life in Capitalist Ruins*. Princeton, NJ: Princeton University Press.

Turner, Jacky. 2006. *Stop-Look-Listen: Recognising the Sentience of Farm Animals*. Hampshire, UK: Compassion in World Farming Trust.

UNEP (United Nations Environment Program). 1992. "Article 2: Use of Terms." Convention on Biological Diversity. Opened for signature June 5, 1992, entered into force December 29, 1993. www.cbd.int/convention/articles/default.shtml?a=cbd-02.

Verspoor, Eric, Lee Stradmeyer, and Jennifer L. Nielsen. 2007. *The Atlantic Salmon: Genetics, Conservation, and Management*. London: Blackwell.

Vialles, Noëlie. 1994. *Animal to Edible*. Cambridge: Cambridge University Press.

Vigne, Jean-Denis. 2011. "The Origins of Animal Domestication and Husbandry: A Major Change in the History of Humanity and the Biosphere." *Comptes Rendus Biologies* 334 (3): 171–81.

Viveiros de Castro, Eduardo. 2005. "Perspectivism and Multinaturalism in Indigenous America." In *The Land Within: Indigenous Territory and the Perception of Environment*, edited by Alexandre Surralles and Pedro García Hierro, 36–74. Copenhagen: IWGIA.

———. 2011. "Zeno and the Art of Anthropology: Of Lies, Beliefs, Paradoxes and Other Truths." *Common Knowledge* 17 (1): 128–145.

Ween, Gro B. 2012. "Resisting the Imminent Death of Wild Salmon." In *Fishing People of the North: Cultures, Economies, and Management Responding to Change*, edited by Courtney Carothers, Keith R. Criddle, Catherine P. Chambers, Paula J. Cullenberg, James A. Fall, Amber H. Himes-Cornell, Jahn Petter Johnsen, Nicole S. Kimball, Charles R. Menzies, and Emilie S. Springer. Fairbanks: University of Alaska, Fairbanks.

Ween, Gro B., and Ben Colombi. 2013. "Two Rivers. The Politics of Wild Salmon, Indigenous Rights and Natural Resource Management." *Sustainability* 5 (2): 478–96. doi: 10.3390/su5020478.

Ween, Gro B., and Marianne E. Lien. 2012. "Decolonization in the Arctic? Nature Practices and Land Rights in the Norwegian High North." *Journal of Rural and Community Development* 7 (1): 93–109.

Winthereik, Brit Ross, and Helen Verran. 2012. "Ethnographic Stories as Generalizations That Intervene." *Science Studies* 25 (1): 37–51.

Zeder, Melinda A. 2012. "Pathways to Animal Domestication." In *Biodiversity in Agriculture: Domestication, Evolution, and Sustainability*, edited by Paul Gepts, Thomas R. Famula, Robert L. Bettinger, Stephen B. Brush, Ardeshir B. Damania, Patrick E. McGuire, and Calvin O. Qualset, 227–59. Cambridge: Cambridge University Press.

Zeder, Melinda A., Ewe Emshiller, Bruce D. Smith, and Daniel G. Bradleys. 2006. "Documenting Domestication: The Intersection of Genetics and Archeology." *Trends in Genetics* 22 (3): 139–55.

INDEX

ability to suffer, 128, 130, 189n1
accountability and care, 129
affective relationality, 61–63
agency: attribution of agency to fish, 61–64; and domestication of fish, 168; and harvest of fish, 133
air and oxygen as preferred medium, 56, 181n8
Alaskan salmon, 122–24
alevins (*yngel*), 88, 108, 111–12; 113*fig.*
alien species designation of farmed fish, 149, 151–53
amoebic gill disease, 39
anchovies, feed pellets and scalability, 120–22, 121*table*
Anderson, David, on relic societies vs. modern man, 14, 176n26
animal-human distinction, 13
animal science and neurological anatomy, 129–33
animal welfare regulations, 141–42, 191nn15–18
Anthropocene epoch, 2
anthropology: anthropological holism, 20, 177n36; domestication in social and cultural anthropology, 11–15; material semiotics of salmon enactment, 19–24, 28–29; and narratives of living, 5–7
Anthropos concept, 12–13, 175n23
Aqkva conference, 27, 28, 30, 40–43
aquaculture: overview, 2–3, 24–25; alien species designation of, 149, 151–53; Aqkva conference, 27, 28, 30, 40–43; breeding programs and Norwegian Red Cattle, 36; contemporary scale of, 4, 173n2; effect on wild stocks, 54–55; employee responsibility and sense of ownership, 79–80, 184n8, 184n11; expansion of, 34–36, 105–7, 106*fig.*, 178n17; as fragile miracle, 73–74; globalization and industry restructuring, 36–37, 39–40; gradation of Atlantic salmon, 78–79, 183n6; hatchery smolt production, 34; history and growth of, 33–35, 40, 46–47; polarized views of, 169–70; production costs, 77–78, 183n1; *rømt oppdrettslaks* (escaped farmed salmon), 149–50, 161–62, 195n30; salmon prices and global spot markets, 78–79, 177n3; in Tasmania, 37–39, 65; wild vs. cultivated salmon, 155, 156*fig.*, 161–62. *See also* domestication; fieldwork; grow-out sites; scalability
Aquanor and Aquasur conferences, 40
Ardener, Edward, on "muted groups," 160–61
Arna River *feilvandring* salmon, 158–60, 194n24
AstroTurf hatchery surfaces, 110, 111–12, 187n8
attachments and scalability, 107–08, 124–25
autopsies, 69–70

Barnes, John, "network" analysis of social class, 44, 47
Bentham, Jeremy, on ability to suffer, 128, 130, 189n1

209

210 INDEX

Berge, Dag Magne, on Scottish vs. Norwegian aquaculture, 35
berggylt (Ballan wrasse), 53–54, 55*fig.*
bevegelige. See sea lice
biomass: overview, 25, 76–77, 102–03; color sampling, 92–95, 95*fig.*; estimation of, 91–96; fish size and sorting, 90–91, 135–36; Fishtalk software, 87–88; government regulation of, 41–42; grow-out phase, 91; *hi sio* ("the other side"; head office), 84–85, 96–100; sale and export, 100–102; and salmon as global bulk commodity, 77–80; sample monthly reports, 88–90, 89*table*; slaughter planning, 99–100; smolt feed conversion ratio (FCR) reporting, 85–88, 89*table*, 90–92; smolt production tanks and monitoring tasks, 80–84. *See also* scalability
biosocial becomings, 15–17
Boas, Franz, 11
boundary objects, 86
Bristol Bay, Alaskan salmon, 122–24
Buller, Henry, on affective relationality, 61–63

cages. *See* seawater cages and pens
Canadian genetically modified salmon, 36
Cattle Tracing System, 188n20
center of calculation, 84–85, 185n19
Childe, V. Gordon, 10, 175n18
Chilean aquaculture, 35–36, 39, 43, 78, 106, 179n18, 183n24
choreographies of care, 190n14
coastline design, 43, 179n28
cognitive and behavioral processes, 130–31
color sampling, 92–95, 95*fig.*
Convention for the Protection of Animals for Farming Purposes, 141
cormorant predation on farmed fish, 38
culling of eggs, 109, 187n7

Darwin, Charles, 9, 11
data collection: *daufisk* (dead fish) removal and counting, 37–38, 52–56, 68–71, 134–35; grow-out sites as laboratory, 67–73; sea lice, 71–73, 182n21. *See also* biomass
daufisk (dead fish) removal, 37–38, 52–56, 68–71, 133, 134–35
The Dawn of European Civilization (Childe), 10
daylight and smoltification, 116–19, 117*fig.*
Despret, Vinciane, on "disembodied body" and presence, 62, 181n12
disease concerns: disinfection procedures, 49; pancreatic disease (PD), 51, 53, 66; *sturing* (moping), 69; in Tasmania, 39; use of vaccines and antibiotics, 74, 183n24; veterinary consultations, 68–70, 73. *See also* sea lice
"disembodied body" and presence, 62, 181n12
domestication: alien species designation of farmed fish, 149, 151–53; in archeology, 10–11; attribution of agency to fish, 62–64; in biology, 9–10; definitions of, 7–9, 166, 174nn9–10; family dogs, 59–60; and farmed salmon, 4–5, 26, 59, 60–62, 164–67, 169–70; as generative process, 166; and human social relationships, 12; and narratives of living, 5–7; and Neolithic Revolution, 8, 174n11; and sentience, 146–47, 167–69; in social and cultural anthropology, 11–15; and unintentional transformations, 170–71
The Domestication of Europe (Hodder), 10
driftsoverskudd (profits), 55
du ser det kokje (boiling on the surface), 66
Dyrevelferdsloven (animal welfare law), 141

economic FCR, 85–86
eggs, 108–11, 179n20
equality as sameness, 46–47
estimation of biomass, 91–96
ethics of care, 128–29
ethnography: biosocial becomings, 15–17; and cospecies histories, 1–2; domestication and sentience, 167–69; material semiotics of salmon enactment, 19–24, 28–29; and modes of thought on farmed salmon, 5, 173n4; salmon ethnographies, 28–29; tracing salmon through people and collaborative fieldwork, 31–32. *See also* growing salmon ethnographies
ettåring (one-year-olds), 117
European Food Safety Authority (EFSA), 139, 141
Evans-Pritchard, Edward, on symbiosis, 12
evolutionary journey of animal domestication, 6, 174n7

farmed fish. *See* aquaculture
feed conversion ratio (FCR) reporting, 85–88, 89*table*, 90–92, 185n23
feeding: and *daufisk* data, 68–71; feed pellets and scalability, 119–22, 121*table*; first feeding, 111–16, 113*fig.*; fish scraps, 45; *gå på svelt*

(go on hunger), 95; *sjekke foringa* (check on the feeding), 57, 64–67, 68–69; at Vidarøy, 49, 50–51, 60–61, 64–67, 75
feilvandring (wandering astray), 42, 158–60
fieldwork: collaborative nature of, 18, 32; combined STS and material semiotics approach to, 17–19, 22–24; language of, 21–22; methods, 29–32; mobile examination of multiple hatchery sites, 29–30, 177n6; patches of relative coherence, 31; practices, 17–19; and researcher connections with topic of interest, 29; scientific sampling and data collection, 71–73; Sjølaks facilities, 30–31; tracking salmon construction sites, 27–28
fin-clipped fish, 148, 155, 156*fig.*, 158–59, 160
first feeding, 111–16, 113*fig.*
fish size and sorting, 90–91, 135–36
Fishtalk software, 87–88
fiskarbonden (fisher-farmer), 35
følgja med på fisken (keep an eye on the fish), 44
forbløffelse (amazement) and spirit of adventure, 21, 177n37
fredløs (outlaw) salmon, 162, 195n29
Frøystad: smoltification, 116–19, 117*fig.*; smolt production tanks and monitoring tasks, 80–84

gå på svelt (go on hunger), 95
gender: ideal of gender equality and opportunities, 46–47; masculine environment of aquaculture at conferences, 42–43; women in Scandinavian aquaculture, 46–47
genetic modification (GMO) ban in Norway, 36
genetics and inbreeding of salmon, 41
globalization and restructuring of aquaculture industry, 36–37, 39–40
Godelier, Maurice, on animal-human distinction, 13
growing salmon ethnographies: eggs, 108–11; feed pellets, 119–22, 121*table*; first feeding, 111–16, 113*fig.*; smoltification, 116–19, 117*fig.*
grow-out sites: overview, 25, 48, 49; affordance, boundaries, and interfaces of, 56–59, 58*fig.*, 181n9; and biomass, 91; *daufisk* (dead fish) removal and counting, 37–38, 52–56, 68–71, 133, 134–35; feeding, 49, 50–51, 60–61, 64–67, 75; and "fragile miracle" of salmon farming, 73–75; jumping salmon, 58*fig.*; as laboratory, 67–73; netting, 57–59; seawater cages and pens, 30, 34, 58*fig.*; *sjekke foringa* (check on the feeding), 57, 64–67, 68–69, 119; smoltification, 30; staff, 48, 51, 54–55, 64–67, 71, 135, 137–38, 139–40, 143–44; Vidarøy facilities, 48–51, 50*fig.*; view from gantry, 68*fig.*; water medium of, 56–57; water surface as relational interface, 59–67. *See also* disease concerns; sentience and salmon harvest
Gulatingsloven, 33
Gullestad, Marianne: on equality as sameness, 46–47; on *ro og fred* (peace and quiet), 180n31
Gyrodactylus salaris, 133

haddock *fiskekaker*, 45
Haraway, Donna: on accountability and care, 129; on mutual ability to share pain, 131–32, 144
Hardangerfjord: Aqkva conference, 41; described, 27; fieldwork at Sjølaks facilities, 30–31; Rebecca's patch in Hardanger, 43–47; sharing of best practices, 36; Vidarøy facilities, 48–51
harvest. *See* sentience and salmon harvest
Hastrup, Frida, on inexhaustible nature of social world, 162
hatchery as heterogeneous construction, 110
hi sio ("the other side"; head office), 84–85, 96–100
history: breeding programs and Norwegian Red Cattle, 36; globalization and restructuring of aquaculture industry, 36–37, 39–40; hatchery smolt production, 34; Norwegian aquaculture, 33–34; politics of Norwegian aquaculture, 34–36; of salmon, 33–37
hoa (mobile females), 71
Hodder, Ian, 10–11, 175n9
høstsmolt (autumn smolts), 118

ideal of gender equality, 46–47
Idunvik hatchery: eggs, 108–10; first feeding, 111–15, 113*fig.*
ikkje ein, men mange (not one, but many), 41
ikkje stresse fisken (not stressing the fish), 45
inexhaustible nature of social world, 162
infectious pancreatic necrosis (IPN), 69, 182n20
infectious salmon anemia (ISA), 39, 183n24
Ingold, Tim: on domestication and human social relationships, 12; on multiple trails of becoming, 15; on unfinished tapestry of life, 163; on wayfaring, 129

Laika (family dog), 59–60
laks (salmon), 33
Law, John: on choreographies of care, 190n14; collaborative fieldwork with, 18, 32; ethnographic study of salmon, 28; European colonial expansion, 111; on "fragile miracle" of salmon farming, 73–74; on material semiotics, 22; on "one-world world," 158, 163
Leach, Helen, on human domestication and evolution, 11
legislation regarding sentience, 140–42, 190n15
leppefisk, 52–53, 181n3
life cycle. *See* grow-out sites
Loke and salmon mythology, 33, 178n10
lubbesild (herring dish), 45
Lusalaus project, 72, 182n22
Lysø committee, 35, 178n14

Marine Harvest, 36
markets as battlefields, 176n35
masculine environment of aquaculture: at conferences, 42–43; women in Scandinavian aquaculture, 46–47
material semiotics: anthropological material semiotics vs. science and technology studies, 32; of salmon enactment, 19–24, 28–29
maximum allowable biomass (MAB), 87, 97, 99, 186n36
Midgley, Mary, on ethics of care, 128
Mol, Annemarie: on ethics of care, 129; on material semiotics, 22
"moral circle" and animal sentience, 127, 129
Mowinckel, Thor, 34
multiple trails of becoming, 15
"muted groups," 160–61
mutual ability to share pain, 131–32, 144
mythology and salmon, 33

Namsen River, 148–49
necropsies. *See* autopsies
Neolithic Revolution, 8, 167, 171, 174n11, 175n17
"network" analysis of social class in a society valuing equality, 44, 47
neurological anatomy, 129–33
"Newcomers to the Farm" project, 18, 176n33
nineteenth-century aquaculture in Norway, 34
Norsk Fiskeoppdrett magazine, 39
Norsk Rødt Fe (Norwegian Red Cattle), 36
Norwegian Biodiversity Information Center (NBIC), 151–53

Norwegian Black List, 151–53
Norwegian Institute of Marine Research, 151–53
"noxious stimuli" vs. pain, 130
The Nuer (Evans-Pritchard), 12
nullåring (zero-year-olds), 117–18

"one-world world," 158, 163
ontological choreographies, 189n6
"Ordinary" salmon, 78, 183n6
outfeed percentage, 68–69
oxygenation and fish stress, 81–82, 185n16

pancreatic disease (PD), 51, 53, 66, 68–70
patches of relative coherence: centers and peripheries of global aquaculture, 39–40; fieldwork, 31
pens. *See* seawater cages and pens
Peruvian anchovies, feed pellets and scalability, 120–22, 121*table*
philosophical views on sentience, 128–29
photoperiod and smoltification, 116–19, 117*fig.*
politics of Norwegian aquaculture, 34–36
pollock, 57–58, 181n10
practical ontology, 23
"Production" salmon, 78
profit margins and global salmon market, 77–80

Regan, Tom, on respect for all "subjects of a life," 128
regulation: industry deregulation and restructuring, 37; legislation regarding sentience, 140–42; Lysø committee, 35, 178n14; Norwegian salmon farms, 46, 180n34; and social practices in Hardangerfjord, 36; sustainable growth goals, 41–42; Viking era *Gulatingsloven*, 33; welfare and accountability, 141nn15–18, 142–45
Regulation on the Protection of Animals at the Time of Killing, 141–42
relic societies vs. modern man, 14, 176n26. *See also* Anderson, David
Research Council of Norway, 40
research methods: anthropological material semiotics vs. science and technology studies, 32; and biosocial becomings, 15–17; fieldwork practices, 17–19; performativity and material semiotics, 19–24
river salmon: overview, 26, 148–49, 162–63; alien species designation of farmed fish, 149, 151–53; Arna River *feilvandring* salmon,

158–60; effect of farmed fish on wild stocks, 54–55; salmon returns vs. salmon escapees, 149–51, 192n7; *smolstskrue* (smolt screw), 156, 157*fig*.; Vosso River salmon rescue project, 153–58; wild, cultivated, and escaped farmed salmon, 155, 156*fig*., 161–62

Rollin, Bernard, on species-specific basis for moral respect, 128

rømt oppdrettslaks (escaped farmed salmon), 149–50, 161–62, 195n30

ro og fred (peace and quiet), 46, 180n31

Rose, James, on "noxious stimuli" vs. pain, 130

salmon: ethnographies of, 28–29; history of, 33–37; life cycle, 30, 33, 177n2. *See also* ethnography; growing salmon ethnographies

salmon farming. *See* aquaculture

scalability: overview, 25, 104–05, 124–25; Alaskan salmon, 122–24; Anna Tsing on, 6–7, 106, 111, 174n8; attachments and scalability, 107–8, 124–25; eggs, 108–11; expansion of salmon aquaculture, 105–7, 106*fig*.; feed pellets, 119–22, 121*table*; first feeding, 111–16, 113*fig*.; smoltification, 116–19, 117*fig*.

scale sampling, 148, 156*fig*.

Scott, James, on strategic barbarism, 14, 176n25

Scottish aquaculture, 35, 62, 178n16

sea lice: and aquaculture regulation, 41; and *berggylt* (Ballan wrasse), 53–54, 55*fig*.; and *bevegelige* (mobile males), 71; data collection, 71–73, 182n21; and *leppefisk*, 52–53; linked to aquaculture pens, 30; mandatory treatments against, 72–73, 182n23; natural remedies, 45, 180n29; scientific sampling and data collection, 71–73; and veterinary consultations, 68–69

seal predation on farmed salmon, 37–38

seawater cages and pens: described, 30, 34; at Rebecca's patch, 44–45, 47; at Vidarøy, 49

sentience and salmon harvest: overview, 26, 126–28, 145–47; animal science and neurological anatomy, 129–33, 189n4; cold water deaths, 136–38; *daufisk* (dead fish) removal, 133, 134–35; and domestication, 146–47, 167–69; forms of death, 132–33; legislation regarding sentience, 140–42, 190n15; philosophical views and, 128–29; salmon as both biomass and sentient beings, 2; slaughtering process, 138–40, 191n17; sorting and separation of fish, 90–91, 135–36; at Vidarøy, 51; welfare and accountability, 142–45

Singer, Peter: on moral circle, 127; on seeking the greatest happiness for the greatest number, 128

sjekke foringa (to check on the feeding), 57, 64–67, 68–69, 119

Sjølaks facilities, 30–31

Skjervold, Harald: breeding of Norwegian Red Cattle, 36; and Tasmanian aquaculture, 38

skjul (cover), 54

småfisk (parr), 90

smolt: feed conversion ratio (FCR) reporting, 85–87; *smolstskrue* (smolt screw), 156, 157*fig*.; smoltification, 30, 116–19, 117*fig*.; smolt production tanks and monitoring tasks, 80–84

steered regime, smoltification, 118–19, 188n19

strategic barbarism, 14, 176n25

Strathern, Marilyn: on different forms of relations, 104; on ideas and thought, 5, 173n3

stress: avoidance of fish handling, 45–46; oxygen saturation levels, 82

studies of technology and science (STS), 18, 23, 32, 111

sturing (moping), 69, 182n19

Sunndalsøra research station, 36

"Superior" salmon, 78–79, 183n6

symbiosis, 12

taming, 9, 175nn15–16

tanks, smolt production, 83–84

Tasmanian aquaculture, 34, 37–39, 65, 87, 106

tromling (net cleaning), 58, 59, 74, 75

Tsing, Anna: on being alive, 176n29; on design and the future, 58; on pixels, 118–19; on scalability, 6–7, 106, 111, 174n8; on worlding exercise, 129, 189n3

ufesk (nonfish), 148, 192n1

unfinished tapestry of life, 163

United Nations Food and Agriculture (FAO) global aquaculture recommendations, 37

vaccination: sorting and separation of fish, 90–91, 135–36, 186n28, 190n7; and use of antibiotics, 74, 183n24

"value adding" and farmed salmon, 78, 105, 187n1

vårsmolt (spring smolts), 117
veterinary consultations, 68–70, 73, 136–37, 144–45, 182n18
Vidarøy facilities. *See* grow-out sites
Vik brothers' development of seawater cages, 34
Viking era and salmon, 33
Vosso River salmon rescue project, 153–58, 193n12, 194n19

wayfaring, 129
wild salmon: and local fishing moratoriums, 54; in Norway, 154, 193n13; and sea lice, 41, 72
worlding exercise, 129, 189n3
worldwide aquaculture expansion, 37
wrasse, *berggylt* (Ballan wrasse), 53–54, 55*fig.*, 181n3

yngel (alevins), 88, 108, 111–12, 113*fig.*

CALIFORNIA STUDIES IN FOOD AND CULTURE
Darra Goldstein, Editor

1. *Dangerous Tastes: The Story of Spices*, by Andrew Dalby
2. *Eating Right in the Renaissance*, by Ken Albala
3. *Food Politics: How the Food Industry Influences Nutrition and Health*, by Marion Nestle
4. *Camembert: A National Myth*, by Pierre Boisard
5. *Safe Food: The Politics of Food Safety*, by Marion Nestle
6. *Eating Apes*, by Dale Peterson
7. *Revolution at the Table: The Transformation of the American Diet*, by Harvey Levenstein
8. *Paradox of Plenty: A Social History of Eating in Modern America*, by Harvey Levenstein
9. *Encarnación's Kitchen: Mexican Recipes from Nineteenth-Century California: Selections from Encarnación Pinedo's El cocinero español*, by Encarnación Pinedo, edited and translated by Dan Strehl, with an essay by Victor Valle
10. *Zinfandel: A History of a Grape and Its Wine*, by Charles L. Sullivan, with a foreword by Paul Draper
11. *Tsukiji: The Fish Market at the Center of the World*, by Theodore C. Bestor
12. *Born Again Bodies: Flesh and Spirit in American Christianity*, by R. Marie Griffith
13. *Our Overweight Children: What Parents, Schools, and Communities Can Do to Control the Fatness Epidemic*, by Sharron Dalton
14. *The Art of Cooking: The First Modern Cookery Book*, by The Eminent Maestro Martino of Como, edited and with an introduction by Luigi Ballerini, translated and annotated by Jeremy Parzen, and with fifty modernized recipes by Stefania Barzini
15. *The Queen of Fats: Why Omega-3s Were Removed from the Western Diet and What We Can Do to Replace Them*, by Susan Allport
16. *Meals to Come: A History of the Future of Food*, by Warren Belasco
17. *The Spice Route: A History*, by John Keay
18. *Medieval Cuisine of the Islamic World: A Concise History with 174 Recipes*, by Lilia Zaouali, translated by M. B. DeBevoise, with a foreword by Charles Perry
19. *Arranging the Meal: A History of Table Service in France*, by Jean-Louis Flandrin, translated by Julie E. Johnson, with Sylvie and Antonio Roder; with a foreword to the English-language edition by Beatrice Fink
20. *The Taste of Place: A Cultural Journey into Terroir*, by Amy B. Trubek
21. *Food: The History of Taste*, edited by Paul Freedman
22. *M. F. K. Fisher among the Pots and Pans: Celebrating Her Kitchens*, by Joan Reardon, with a foreword by Amanda Hesser
23. *Cooking: The Quintessential Art*, by Hervé This and Pierre Gagnaire, translated by M. B. DeBevoise
24. *Perfection Salad: Women and Cooking at the Turn of the Century*, by Laura Shapiro

25. *Of Sugar and Snow: A History of Ice Cream Making*, by Jeri Quinzio
26. *Encyclopedia of Pasta*, by Oretta Zanini De Vita, translated by Maureen B. Fant, with a foreword by Carol Field
27. *Tastes and Temptations: Food and Art in Renaissance Italy*, by John Varriano
28. *Free for All: Fixing School Food in America*, by Janet Poppendieck
29. *Breaking Bread: Recipes and Stories from Immigrant Kitchens*, by Lynne Christy Anderson, with a foreword by Corby Kummer
30. *Culinary Ephemera: An Illustrated History*, by William Woys Weaver
31. *Eating Mud Crabs in Kandahar: Stories of Food during Wartime by the World's Leading Correspondents*, edited by Matt McAllester
32. *Weighing In: Obesity, Food Justice, and the Limits of Capitalism*, by Julie Guthman
33. *Why Calories Count: From Science to Politics*, by Marion Nestle and Malden Nesheim
34. *Curried Cultures: Globalization, Food, and South Asia*, edited by Krishnendu Ray and Tulasi Srinivas
35. *The Cookbook Library: Four Centuries of the Cooks, Writers, and Recipes That Made the Modern Cookbook*, by Anne Willan, with Mark Cherniavsky and Kyri Claflin
36. *Coffee Life in Japan*, by Merry White
37. *American Tuna: The Rise and Fall of an Improbable Food*, by Andrew F. Smith
38. *A Feast of Weeds: A Literary Guide to Foraging and Cooking Wild Edible Plants*, by Luigi Ballerini, translated by Gianpiero W. Doebler, with recipes by Ada De Santis and illustrations by Giuliano Della Casa
39. *The Philosophy of Food*, by David M. Kaplan
40. *Beyond Hummus and Falafel: Social and Political Aspects of Palestinian Food in Israel*, by Liora Gvion, translated by David Wesley and Elana Wesley
41. *The Life of Cheese: Crafting Food and Value in America*, by Heather Paxson
42. *Popes, Peasants, and Shepherds: Recipes and Lore from Rome and Lazio*, by Oretta Zanini De Vita, translated by Maureen B. Fant, foreword by Ernesto Di Renzo
43. *Cuisine and Empire: Cooking in World History*, by Rachel Laudan
44. *Inside the California Food Revolution: Thirty Years That Changed Our Culinary Consciousness*, by Joyce Goldstein, with Dore Brown
45. *Cumin, Camels, and Caravans: A Spice Odyssey*, by Gary Paul Nabhan
46. *Balancing on a Planet: The Future of Food and Agriculture*, by David A. Cleveland
47. *The Darjeeling Distinction: Labor and Justice on Fair-Trade Tea Plantations in India*, by Sarah Besky
48. *How the Other Half Ate: A History of Working-Class Meals at the Turn of the Century*, by Katherine Leonard Turner
49. *The Untold History of Ramen: How Political Crisis in Japan Spawned a Global Food Craze*, by George Solt

50. *Word of Mouth: What We Talk About When We Talk About Food*, by Priscilla Parkhurst Ferguson
51. *Inventing Baby Food: Taste, Health, and the Industrialization of the American Diet*, by Amy Bentley
52. *Secrets from the Greek Kitchen: Cooking, Skill, and Everyday Life on an Aegean Island*, by David E. Sutton
53. *Breadlines Knee-Deep in Wheat: Food Assistance in the Great Depression*, by Janet Poppendieck
54. *Tasting French Terroir: The History of an Idea*, by Thomas Parker
55. *Becoming Salmon: Acquaculture and the Domestication of a Fish*, by Marianne Elisabeth Lien